WORKSPACE
STRATEGIES

WORKSPACE STRATEGIES

Environment as
a Tool for Work

Jacqueline C. Vischer

Buildings-in-Use, Boston and Montreal

CHAPMAN & HALL

New York • Albany • Bonn • Boston • Cincinnati • Detroit • London • Madrid • Melbourne
Mexico City • Pacific Grove • Paris • San Francisco • Singapore • Tokyo • Toronto • Washington

Cover design: Curtis Tow

Copyright © 1996 by Chapman & Hall

Printed in the United States of America

Chapman & Hall
115 Fifth Avenue
New York, NY 10003

Chapman & Hall
2-6 Boundary Row
London SE1 8HN
England

Thomas Nelson Australia
102 Dodds Street
South Melbourne, 3205
Victoria, Australia

Chapman & Hall GmbH
Postfach 100 263
D-69442 Weinheim
Germany

Nelson Canada
1120 Birchmount Road
Scarborough, Ontario
Canada M1K 5G4

International Thomson Publishing Asia
221 Henderson Road #05-10
Henderson Building
Singapore 0315

International Thomson Editores
Campos Eliseos 385, Piso 7
Col. Polanco
11560 Mexico D.F. Mexico

International Thomson Publishing–Japan
Hirakawacho-cho Kyowa Building, 3F
1-2-1 Hirakawacho-cho
Chiyoda-ku, 102 Tokyo
Japan

1 2 3 4 5 6 7 8 9 10 XXX 01 00 99 98 97 96

Library of Congress Cataloging-in-Publication Data

Vischer, Jacqueline.
    Workspace strategies : environment as a tool for work / by
Jacqueline C. Vischer.
        p.   cm.
    Includes index.
    ISBN   0-412-07411-7 (alk. paper)
    1. Work environment.   2. Offices.   3. Buildings—Environmental
engineering.   I. Title.
T59.77.V59   1996
658.2'8--dc20   96-18623
                                                                      CIP

To order this or any other Chapman & Hall book, please contact **International Thomson Publishing, 7625 Empire Drive, Florence, KY 41042.** Phone: (606) 525-6600 or 1-800-842-3636. Fax: (606) 525-7778. e-mail: order@chaphall.com.

For a complete listing of Chapman & Hall titles, send your request to **Chapman & Hall, Dept. BC, 115 Fifth Avenue, New York, NY 10003.**

This book is dedicated to Isabelle Claire Vischer Skaburskis, and her friends.

CONTENTS

FOREWORD

We live in era of transformation—of technology, of social values, and of the way work is done. This book represents a timely and innovative addition to current thinking and writing about transformation in organizations.

In order to meet an increasingly global and competitive environment, organizations are undergoing reengineering, work process redesign, "right sizing," creating a "virtual office," and other forms of restructuring and basic change of the way work is accomplished. Such transformation means analyzing and redesigning core processes in organizations around new kinds of principles such as "total quality" and customer service. The eventual effect of these changes is likely to be the networked or "boundary-less" organization, in which the traditional boundaries between functions and between producers and their suppliers—and sometimes even between organizations and their competitors—are broken down. The goal of such transformation is to make the work of the organization more efficient and productive—to produce more with fewer resources and at a lower cost.

In the conventional view of the transformation process, certain secondary concerns, such as the need to protect the environment or to help an increasingly heterogeneous work force deal with its personal issues, are seen as problematic for this core thrust. Some recent work, however, is beginning to show that if these so-called secondary concerns are considered central, far from being problematic, they actually present strategy opportunities for productive innovation and change.

So, for example, changing core production techniques for the sake of actually decreasing the output of waste (instead of emphasizing only waste disposal) can lead, often unexpectedly, to more efficient and cost-effective production. Similarly, looking at work through the lens of employees' work/family issues can also, again unexpectedly, lead to work practice innovations that not only help employees but actually increase productivity. In other words, if secondary issues are seen as opportunities for learning and strategic change, rather than as problems to be dealt with separately and marginally, they themselves can contribute to the goals of work-place transformation.

To this burgeoning view Jacqueline Vischer has added another intriguing dimension: the role of the physical space within which employees work. In her book, she has numerous examples that show the limitations of treating space planning and space-related issues without concern for the potential strategic value of space decisions. She demonstrates the critical import that details of workspace design have on human performance. Her book provides an innovative and efficient way in which workspace planning, design, and evaluation can be linked to business planning and business objectives, and thus strengthen core goals. The detailed case studies presented in this book document the strategic advantages of a more central approach to this aspect of the workplace, and show how approach can be implemented in organizations.

Jacqueline Vischer's book is important for the productive management of workplace accommodation. It makes amply clear that when accommodation is looked at strategically, both workers and organizations gain. It also contributes significantly to a more holistic view of organizational processes, and to an emerging understanding that keeping a boundary between primary and secondary concerns is no longer useful. By providing a critical and strategic understanding of workspace, the book presents a valuable fresh look at the transformations of work currently underway.

Lotte Bailyn[*]
T. Wilson Professor of Management
Massachusetts Institute of Technology
Cambridge, Massachusetts

---

[*]Author of *Breaking the Mold: Women, Men, and Time in the New Corporate World*, Free Press, 1993.

PREFACE

After publishing *Environmental Quality in Offices.* I began to be invited to consult to companies and public agencies on solving problems in office buildings that were considered to be "psychological." One example of a psychological problem are buildings where numerous changes have been made to the air handling systems, but people are still complaining about indoor air quality problems. In another example, workers were actually going out and buying "full-spectrum" fluorescent light bulbs for their floor because they were tired of complaining to management about discomfort from regular, cool, white fluorescents. And in a third, employees in the small branch offices of a bank were reporting ergonomic problems in their necks and hands, and the facilities staff, being used to activities like moving, new construction, and purchasing furniture and carpet, did not know how to respond.

So I formed Buildings-In-Use (and later opened an office in Canada, with the additional name of Bâtiments-en-Usage) and in carrying out BIU surveys for clients with these kinds of problems. I began to learn about the different events that can lead to, and flow from, an activity like Building-In-Use Assessment—that is, asking users for diagnostic information about their workspace. I saw what happens in companies when large amounts of information about space layout and building quality are received, and how important it is to be prepared to handle that information. I was challenged by managers who wanted to be able to do something else with BIU Assessment, something more than acquire diagnostic data on their buildings. Some proposed finding ways to record responses electronically; some wanted to integrate BIU scores with their CAD or CAFM database; some proposed a comparison of the relative comfort of different types of furniture system; and others wanted a satisfaction survey to improve the "image" of facilities managers, and to generate better public relations with building users.

As a result of all these changes and demands, the BIU Assessment system began to stretch and expand. Innovations became possible that had not been thought of when I developed the system at Public Works Canada in the 1980s. From being a good quality standardized survey tool for collecting diagnostic feedback from building users, the BIU Assessment system began to blossom as a tool for user-manager communication, a resource for space planners and designers, and a way for business managers to make space-related decisions.

I also began to learn more about building users themselves—how office workers judge the physical attributes of their workspace, especially those which are ubiquitous in modern office buildings. The fluorescent lighting (whether in coffered ceilings, in recessed fixtures, or just plain flat lights in the ceiling); the acoustic partitions, their height, their color, their effectiveness, their use and misuse, the problems people were likely to have with indoor air temperatures; the concern they felt when there were unexplained odors, or smoke in no-smoking areas; and their feelings about colors, about noise, about the cafeteria, and about space standards and furniture. Over the years that people spend in a job, they can become very attached to their workspace. They can also become astute and sensitive analysts of the impact their space has on their ability to work. I learned that feelings can run high when there is a perceived problem in a building, and more than technical expertise about buildings is needed to survive the politics of "workspace wars".

This book was therefore written in order to organize and present these experiences to readers, to demonstrate not just how BIU Assessment works, but also the impact on an organization of introducing such a system. For most North American an European companies, initiating feedback from users about their workspace is a new departure (although many have introduced customer satisfaction surveys), and it is important to see BIU Assessment not just as a way diagnosing building problems, but also as a movement towards change and the reengineering of work. The great potential of a system like BIU Assessment for an organization is that it provides a handle on environmental quality which can help facilitate business transformation and should in any event accompany the re-engineering of work processes. In this way, the process of acquiring feedback from building occupants fits in with the current movement towards new ways of doing business. The chapters you are about to read will address all these levels of BIU Assessment—the process as well as the product—and will inform you about how people relate to space at work and how it can be made to work better for them.

The book starts out by explaining the *Organization–Accommodation Relationship*—the stages through which an organization moves in its relationship to the spaces and physical environment occupied by its employees. The premise of the early chapters is that in the vast majority of cases, the O–A relationship could be improved—it needs to be better understood by managers and decision-makers, and tools need to be made available to analyze and improve it whenever necessary. The O–

A relationship should serve the needs of the organization, and help it to be successful, rather than simply ignored, or treated as another cost item.

The book draws on examples of recent, more humane, office architecture (mostly in Europe), to demonstrate that a better work environment can be achieved in modern office buildings without necessarily increasing costs or reducing worker productivity. The relationship between the work environment and human behavior, including what is known about productivity, morale, and human comfort, is explored, and the concept of "functional comfort" is introduced: a yardstick for evaluating the degree to which features of the physical environment have an impact on people's performance of work.

The book describes results of numerous research studies that have been carried out on how space design in office buildings affects people at work and draws attention to the fact that neither facilities managers, business managers, nor architectural designers routinely draw on this knowledge to ensure that workspace is designed to optimize the performance of work.

The middle chapters of the book describe the Building-In-Use (BIU) Assessment system—one approach to eliciting feedback from building occupants designed to help decision-makers develop strategies for optimizing workspace. Used by large corporations in the United States and Canada, this feedback system uses a standardized measurement questionnaire and compares seven key building scores to database norms from a 3,000-case database. The system functions diagnostically to indicate priorities for intervention and improvement. A process for implementing BIU Assessment in organizations is outlined in some detail.

These chapters also provide case studies of companies who have used Building-In-Use Assessment to elicit feedback from occupants which they have then used to solve building problems, or to decide which of several competing improvements it is most cost-effective to spend money on. The book enters in some detail into the seven key dimensions of workspace quality measured by BIU Assessment, what they mean, how they affect people, and what solutions to typical environmental problems are encountered in modern office buildings.

The final chapters of the book explain that implementing occupancy feedback systems such as BIU Assessment means empowering employees to understand and take responsibility for their own work environment, and that this implies a major shift in values for many traditional companies. The politics of initiating occupancy feedback are explored in some detail, so as to prepare managers for barriers they might encounter when they try to implement an occupant feedback system in their own companies. The experiences of three companies are explored in depth to show how well they succeeded in establishing a BIU Assessment system within their—very different—organizations. These also demonstrate the considerable payoff to a company of using occupant feedback to improve the O-A relationship.

In the last chapter, Accommodation Planning is linked with Strategic Business Planning to show how theories of strategic business planning can be expanded to take accommodation planning into consideration. The book points out the advantages and cost savings of optimizing accommodation to suit employees' task requirements—not only by reducing occupancy costs and saving space, but also by making employees more productive. The book shows how workspace can and should be designed to function as "a tool for work"—like the computer or the telephone—in order to provide a return on the significant investment a company makes in its space.

ACKNOWLEDGMENTS

Needless to say, this book could not have come into being without the supportive and positive clients of Buildings-In-Use/Bâtiments-en-Usage, in Canada and in the US, who, in hiring us to work for them, demonstrated faith in our concepts and belief that our way of thinking would do good for their companies. It was through working for such clients—learning about their problems, goals, and ways of working, and developing responses to their needs—that the ideas and concepts that form the basis of the Building-In-Use approach were really developed. Each one of our clients—and their individual representatives, too numerous to name here—had something to add to what the BIU system could do for them, and how they wanted it applied. They also had questions and expectations—and it was in finding responses to their needs and answering their questions that BIU developed its strength and rooted its ideas in a context of reality. In am also grateful for the contribution of all BIU staff and associates over the years—especially to Dori Frewald Mock and to Gwen Shiels—for participating in this process of evolution.

I would also like to acknowledge the debates and discussions held in the context of Space Planning in Organizations Research Group (SPORG) at MIT. The members of this group—both core members and visitors—share a vision of buildings, space, design, and human behavior that has developed over years of meetings, has evolved through its research projects, and has been articulated in numerous lively and innovative discussions with people about their work in this field.

I would like to thank Kreon Cyros for inviting me to participate in his courses and conferences and for access to his Facilities Management Library. Special thanks to Carolyn (Gemma) Kerr, Dawna Paton, Tim Springer, and Dale Tiller for reviewing the manuscript in draft and offering invaluable suggestions and corrections, and particularly to John Zeisel, who edited the manuscript and helped me formulate my ideas.

Finally, I would like thank Isabelle and Evan for feeding and caring for themselves uncomplainingly while this book was being written.

Jacqueline C. Vischer PhD

CHAPTER 1

CORPORATE EXCELLENCE IN FACILITIES: WHY THE WORKPLACE IS IMPORTANT

"When effectively linked to a company's business priorities, buildings' design contributes to corporate image, to serving both customers and employees, and can enhance creativity and accelerate the development of products"

Peter Lawrence

THE ORGANIZATION–ACCOMMODATION (O–A) RELATIONSHIP

The relationship between an organization and the physical environment it occupies—its building, or space in a building—is a key dimension of that organization's strength, growth, and success. Companies, government departments, institutions, even families, make emotional, financial, personal, and corporate investments in the O–A relationship, often without being aware of its importance. Increasing awareness of the dynamics of the interaction between building users and their accommodation is a key to improving the quality of this relationship, to the mutual benefit of both sides. How does this relationship work?

A large and well-established property management firm in Boston, Massachusetts, recently retained a law firm to help combat charges of indoor air pollution and "sick building syndrome" in one of their buildings. One of their tenants was complaining, and only one; the other occupants of the building appeared unconcerned. The complaining tenant was locked into a long-term lease at the peak rents that downtown office buildings were demanding a year or so earlier. The year-long downturn in the economy resulted in a large amount of more attractively priced office space on the market. The tenant was dealing with business worries resulting from the economic slowdown and was clearly keen to negotiate its way out of its lease and into cheaper space. The company's employees began reporting headaches, nausea, and respiratory problems in record numbers. Their managers asked them to document the type and frequency of their symptoms. The tenant then used this documentation to support a request to the property management firm to take immediate action to "solve the problem." The property managers, after carrying out extensive investigations of the mechanical systems in the building, could not discover any differences between the floors occupied by this tenant and other areas of the building. They also examined air samples for chemical contaminants and bacteria and looked for evidence of fungi and molds. These tests yielded no evidence of indoor air pollution. But the tenant claimed the building was poisoning its employees and that this was grounds for breaking the lease. The property management firm suspected that by encouraging its employees to write down all its symptoms, the tenant had encouraged a sort of hysteria to develop, and all along had intended to claim "sick building" problems to have an excuse for moving out of the building.

The property management company filed a counter-lawsuit against its former tenant for illegally breaking the lease. The management company had spent over $50,000 testing, repairing and upgrading the ventilation systems, and it was expecting to pay tens of thousands more in legal costs. But as one of their spokespersons pointed out, "One of the worst things you can say about a building is that it has environmental problems. It's like saying a car has faulty brakes."[1] In other words, the long-term costs of managing a building thought to have environmental problems were likely to be far more dramatic than the short-term relatively high costs of salvaging that building's reputation in court.

This story says some important things about the role played by real estate in general and office space in particular in modern corporate life.

- *Dollars:* it is a vivid illustration of the financial burden of the costs of office space for small and medium-sized companies, and to what lengths they might go to reduce that burden.

- *Health*: it demonstrates the power of the "sick building syndrome", and the latent anxiety that all office employees have about their health in sealed office buildings.
- *Management*: it illustrates the need for the high level of technical expertise, interpersonal skills, building knowledge, and diplomacy required by those who manage a modern office building.
- *Business strategy*: it throws into relief the critical role of the O–A relationship in the context of doing business.

An organization's accommodation is therefore about money, health, people and business, and it deserves to be better understood.

An organization that does not fit well into its accommodation is less than effective in carrying out its mission, regardless of whether its discontent arises from occupational health hazards, excessive expenditures, inept management services, or inappropriate space. Accommodation that is functional, appropriate, and cost-effective enhances the performance and the productivity of individual employees and also of the organization as a whole. It is important, therefore, to understand the O–A relationship and to determine ways in which an organization's accommodation can be made to work *for* rather than *against* it over the period of its occupancy.

The O–A relationship is like a marriage. There are identifiable stages in the evolution of the relationship. There are different agents or intervenors at each stage—mostly belonging to the real estate industry—and different decision-makers apply different quality criteria at each stage. [2] There are at least five stages: finding space, planning and design, moving in and settling down, adaptation and change, and moving on, or out. Each of these stages is considered below.

Finding Space

In the first stage of the O–A marriage, when a company is looking for its space, leasing agents, like match-makers, may be involved to find and negotiate office space with a range of conditions, services, and dollar amounts. If the company is large and considering building a building, it will likely study the cost-benefit trade-offs of building versus leasing space. In either case, building appearance and image are important criteria, as the popularity of many grandiose urban office buildings testifies.[3] Studies also show that many decisions, that later turn out to be critical, are dictated by misplaced cost considerations at this stage of the process.[4]

Planning and Design

Once space has been identified, the O–A relationship moves into the second stage. The occupying organization plans its space, maybe constructs it, builds out the interior, and prepares for occupancy. This process can be long and expensive, in part because no one knows what it should cost. In high-priced office buildings, the cost of building out interiors is often borne by the landlord as part of the services provided to incoming tenants. At this stage, a designer or architect is called in, and one or more employees—occasionally a senior executive—are assigned to work with a team of designers to make decisions for the new space.

The designers sometimes work with facilities managers and sometimes with employee representatives to plan interior space lay outs. Sometimes they are assigned to the company's project manager and have no contact with the people who will operate or who will occupy the space. As it is at this stage that the organization determines its needs and mode of operation in order to create a fit between these and the physical environment in which it is to function, the process used may result in a fit between the organization and its accommodation that is good, bad, or indifferent.

Moving In and Settling Down

The third stage of the O–A marriage—moving in and settling down—is also costly, because employees' work is affected by the process of packing and moving, and because inevitably many changes have to be made, some small and some not so small, to help the occupants adjust to their new space. Most building managers estimate that it takes one full year for a new building to settle down and for people to settle in, after which the process of environmental adjustment and change that accompanies occupancy may slow down, although it rarely stops.

As soon as a company moves employees into new space—and even when it doesn't—the process of churn begins. As work-groups are moved, cut, merged, added, and transformed, space changes and adjustments are made on an almost continuous basis. In modern offices, the churn rate ranges from 20 percent to 90 percent per year, with most falling between 30 percent and 50 percent—a costly budget item in terms of one in-house estimate of $1,000 to move one workstation.[5] Churn is an expensive corporate indulgence—one of the common sources of discomfort in the O-A relationship—especially where managers are accommodated in enclosed offices so that walls have to be demounted and rebuilt each time the managers move. Even furniture systems, designed to be easily moveable, can require a team of workers to deconstruct and re-erect the partitions. And when the furniture moves, cabling requires adjustment, telephones have to be coordinated, and whole areas have to be repainted and recarpeted. Sometimes, a planned accommodation change leads to a move being successfully executed, only to reveal that that particular function of the organization is being dissolved, merged with another group, or moved out of the building. Space that is carefully planned for one work-group can rapidly become inefficient and uncomfortable when new equipment is added, new technology introduced, or people are added or removed. In a company seeking to improve its space to help people work, inappropriate accommodation can rapidly become dysfunctional.

Adaptation and Change

During its occupation of a building—the fourth stage of the marriage—specific issues may arise which affect the O–A relationship. For occupants of many new and not-so-new buildings, the process of interacting with their space is more like a battle than a marriage. And as long as employees feel they are fighting their accommodation to get their work done, the more demands they place on building managers to meet their needs. For example, an inadequate ventilation system may give rise to indoor air quality complaints and thermal comfort problems. Building managers of owner-occupied buildings have to decide whether or not to invest in a renovation; tenants wonder whether they can oblige the landlord to pay for modifications when their space is leased. The electric power available may be insufficient if computer equipment is added, and power outages may start to plague the company's computer systems. Parking may turn out to be inadequate, with no possibility of adding space. Or the light fixtures in hallways may be positioned in such a way as to make changing light bulbs costly and difficult.

During occupancy, therefore, and often from the very earliest days, the relationship between an organization and its accommodation fails to improve, and, in fact, deteriorates. A small organization, like a family which occupies a relatively small amount of space, will fix what it can or move elsewhere. However, large companies and institutions do not have such flexibility and continue to spend money on efforts to salvage the O–A relationship. There is little published information on appropriate expenditures for operating and maintaining space to help decision-makers determine what to spend money on, and how much to spend to fix problems in the O–A relationship. [6] Even less is known about the employee efficiency, comfort, and health costs to a company of a bad O–A relationship—that is to say, of occupying space that fails to meet its needs efficiently.

Moving On or Out

If a company is successful and growing, it may eventually, in the final stage of its O–A relationship, decide to move. Or, if it is stable, it may decide to stay where it is. Companies leasing space may decide to renegotiate terms; or, if they own their space, they may renovate and upgrade. If a company is not thriving, it may try to shrink down its space or move to smaller premises. What do companies learn from their O–A experience that will help them make a good decision? In the time that has elapsed between the first stage of the O–A relationship (Finding Space) and now, millions of dollars have been spent on a relationship for which there are no clear measures of success, or failure.

The company probably knows little about the long-term effectiveness of its accommodation expenditures, or about the impact of decisions that were taken at each stage. Was is better to build than to lease space? Was the move coordinator's fee worth the time saved? Did the design decisions hold up once the space was occupied by people at work? Were there more or less than the usual range of fine-tuning problems to solve after move-in? Were the accommodation decisions that were made cost-effective for the firm, and how can the firm find this out?

ADDING VALUE TO THE O–A RELATIONSHIP

When manufacturing industries dominated business, accommodation was termed *plant* and buildings were commonly referred to as *facilities*, which included the machinery and the workers as well as the buildings they occupied. Corporate business strategy defined facilities in terms of their production capacity, and changes (usually expansion of space and/or equipment) were addressed through the capital budgeting process as Bower has pointed out.[7] In industries such as automobile and consumer goods manufacturing, plants were periodically closed and redesigned for new equipment or processes.

The office building, however, has no such cycle. It is in a constant state of accommodating change. In this, as in its role as an active element in the performance of work, the white collar work environment is handled less effectively than its manufacturing counterpart. As our economy moves towards the predominance of white-collar work, it seems apparent that the manufacturing model of planning and acquiring space has not been effectively applied to business decisions about office accommodation and space for office tasks and white-collar workers.

In an early definition of the cooperative dynamic that constitutes an organization and its structure and functions, Barnard recognized the inherent role of physical space in the functioning of a business:

> An inspection of the concrete operations of any cooperative system shows at once that the physical environment is an inseparable part of it.... That part of the organization...which consists of structures, improvements, tools, machines, etc. pertains to the organization which owns or works with them. For this reason, in many cases the notion of an organization evidently includes that of a physical plant; for example, in the case of a railway or a telephone organization. It is apparent that when one is dealing with a specific enterprise the whole situation comprising physical plant, men and activities must be the minimum system with which one is primarily concerned.[8]

He goes on to distinguish between an *enterprise, business* or *operation*, and the term *organization* which is "reserved for that part of the cooperative system from which physical environment has been abstracted." He adds,

> All aspects of the physical environment are then regarded or most conveniently treated as the elements of other, physical and technical, systems, between which and organizations the significant relationships may be investigated as may be required for the purpose in hand.[10]

Clearly, one critical purpose for investigating such relationships is to guide and inform business strategy. In the context of manufacturing, investment in physical plant was and is an important element of business strategy because of the amounts of capital involved. Deciding how much to spend on plant expansion and equipment is strategically calculated on the basis of value creation, that is to say, capital budgeting decisions on which excess return can be anticipated.

With the reductions in the manufacturing sector and the upsurge in white-collar work, Barnard's way of thinking about the physical environment has been neglected in the present-day business environment. This hiatus has created a strategic gap in today's corporate and business planning. Barnard's reference to railway and telephone companies was made at a time when huge capital investments were needed in physical infrastructure by both these industries. While this form of investment in plant and facilities is no longer taking place, investment in communications and computer technology has replaced it. Telephone companies offer a good example.

As communications technology has developed, increasingly powerful switching units occupy decreasing amounts of space. Facilities that were built to house transmission equipment 15–20 years ago are now half empty, yet carrying far more lines. With the advent of cellular and fiber-optic technology, the space required for physical infrastructure will dwindle even more, but this is not to say that telephone (or as they are now called, telecommunications) companies do not need facilities. It is just that their facilities are more likely to be filled with people than equipment: telephone operators (although their number is shrinking), customer service representatives, information services, accounting and administrative staff, and, growing by far the quickest in an increasingly competitive environment, sales and marketing representatives. Even railway companies are adding faster to their white-collar work force than to their physical infrastructure, using computerized reservation and routing systems, for example, operated by clerks at desks in offices.

Fewer telecommunications personnel operate large, complex, and expensive machinery than in the past: an ever growing and highly technical system requires fewer operators. Telecommunications workers rely on sophisticated electronic equipment to communicate as they travel to see clients, to access client databases through networked computers, and to process orders and billings through a series of communicating computer systems. They are trained on and can access a large number of specialized telecommunications services that they both market to clients and use themselves to monitor revenues, costs, billing, and client behavior.

The telecommunications business illustrates the vast difference between the strategic role of capital budgeting decisions a generation ago and the strategic role of capital budgeting decisions for today and tomorrow. What used to be investment in physical plant and in equipment, including the space to accommodate it and the people operating it, has been superseded in many of today's businesses by investment of capital funds in office space, furniture systems, and electronic and communications technology. Separate organizational units such as facilities management, space planning, and information systems groups have grown up to handle these specialized areas of operation, thus moving these functions out of the business units and out of the realm of operational decision making—a critical step, as a result of which the key link between business planning and capital spending has weakened. Whereas Bower can conclude that "the same [operating managers of a business] or their successors are the ones who implement the business plan on which the capital proposal is based", evidence from the contemporary O–A relationship suggests that today's white-collar business managers are not giving their accommodation the same detailed attention as their manufacturing predecessors gave to their capital expenditures on plant and facilities.

THE UNEXAMINED O–A RELATIONSHIP IS NOT WORTH HAVING

Consider the stories of two banks. The first, a large New York bank, bought out a Boston bank and determined that new premises were needed in the Boston region for its new enterprise. The facilities manager of the Boston bank was put in charge of this process, which was far beyond his usual range of responsibility. So he retained some architects to determine an appropriate building in which some of the Bank's departments might lease space and to design the space they were to occupy. He managed the selection process by requesting bids from local architecture firms, devising a short list of applicants, and selecting the most cost-effective proposal in terms of amount of services offered for dollars charged. The architects determined which were suitable buildings in Boston for the people employed in the bank's business units and began to design their office space. At the time when the bank's directors approved the selection of one of the buildings, important points were being decided regarding the company's space needs. It became apparent that no one actually knew how much space was needed, because no one had been told exactly which work-groups were to be accommodated in the new building. Some employees were going to lose their jobs as a result of the merger, and others would be moved out of the city into decentralized bank operations in the suburbs. But this information was not available to the architects who assured their client they did not need it. Their architects' mandate was simply to select a building and to design the space. The facilities manager saw his mandate as moving into the new building with a minimum of demands on senior management. His job, as he saw it, was not to ask a lot of difficult questions about relocating numbers of people, but to do the job as quickly and efficiently as possible. He did not want his new bosses to think he was not competent to manage the project, so he did not demand answers to the questions of who was to be accommodated—or where—from his superiors. The architects, for their part, did not want to appear unresponsive to the client, so they did not ask any questions about the users of the space they were designing. After all, they said, office space is pretty much the same wherever you go.

The process experienced by the bank's business units is typically the sort of sequence of reactive and unplanned decisions which results in a work environment with which the organization has the usual poor and costly O–A relationship—and they may not even be aware of it. Employees find themselves accommodated in too little or too much space, and work-groups are moved out almost immediately after moving in. No matter how good the quality of the building, unanticipated equipment loads generate thermal comfort problems and excessive demands on the electrical and mechanical systems. People feel uncomfortable with ventilation, thermal comfort, and lighting, but do not protest too loudly because they are glad to have kept their jobs. There is a long settling-in period, with the facilities managers keeping busy fixing building problems and adjusting equipment. The organization's managers have no way of knowing the costs of this problem-fixing process, or the longer-term costs of the accommodation decisions that were made. But for them, accommodation is written off as overhead. They will not explore how much money they could have saved by making better accommodation decisions, or by integrating the accommodation planning process with their business strategy. Shareholders have even less way of knowing how much these kinds of oversights might have eaten into profits. No one will know to what extent the office interiors slow down employees in their work, cause low morale and absenteeism, and have a negative effect on their relationship with customers. No one, in effect, ever systematically calculates the costs to an organization of its poorly-made accommodation decisions over the time it occupies its office space.

Compare this story with that of another bank: the Bank of Boston. This company reduced operating losses and overhead costs while investing in the quality of the O–A relationship and adding value to the firm's products.[12] This bank began its process by seeking ways to improve the profitability of certain business units which were not competitive. A close examination of these operations showed that work processes were not efficient, that several groups replicated each other's tasks, and that key personnel were leaving because of the uncomfortable physical environment in which they were working. A series of strategic business decisions which included updated and improved software, a shift to "just-in-time" work processes for which all employees, management and staff, received training, and consolidation of seven different work-groups under one roof, led to the decision to move all employees into a new building. The selection of the building (located with access to demographically appropriate regions for the bank's labor needs), and design of the interior (detailed to correspond to and facilitate the just-in-time procedures) constituted a series of decisions made by the bank's business unit directors; the process was guided by, but never relegated to facilities or design staff.

Figure 1.1. "Just-In-Time" office in the Bank of Boston, Canton, Massachusetts. Photo: Turid Horgen.

After considerable searching, and having decided not to build, the bank selected a building to lease in a town south of Boston. This building required extensive interior fit-out to accommodate the regrouped business units, and designers were hired to design the building's interior while consultants worked with employees to streamline their workflow processes. The new offices incorporated innovative design solutions to issues of work-flow efficiency and flexible work-group planning; it cost some nine million dollars to build out. After move-in, the improved efficiency of work-group operations resulted in a 30 percent reduction in the amount of space required to perform the same amount of work. The bank was also able to reduce the number of employees by 25 percent. The space vacated in the new building has been occupied by another work-group, saving money on premises leased elsewhere by the bank. The overall savings to the company in space and payroll terms was well over $9,000,000 within two years of occupancy. More-over, in spite of the fears they expressed about the move, employees like working in the new building, and the bank has been able to use its new facility as a showpiece to attract new clients. This strategy has thereby further increased the profits of the business units located on these premises.[13]

Table 1.1 summarizes the contrasting examples of O–A relationships in a contemporary business environment.

Table 1.1.   Contrasting Examples of O–A Relationships in a Contemporary Business Environment

The unexamined and more typical O–A relationship	The more positive and value-added O–A relationship
The physical environment in which a company is accommodated is at best a compromise, most likely a disappoiment, and, at worst, a liability.	The physical environment in which employees are accommodated is cost-efficient to operate and conducive to the performance of work.
The physical environment in which a company is accommodated will almost certainly deteriorate over time because it is expensive to continually renovate.	The physical environment in which people work adapts to the changing needs of individuals and groups over time because renovations and upgrades are an investment in staff performance
The lack of corporate interest in the process results in dependence on an array of outside actors and agents who are often not concerned for or knowledgeable about the business strategy of the firm.	The industry of real estate experts can be counted on to respond knowledgeably and responsibly to clients' needs because decision-makers are involved, and articulate about what they want.
Company executives rarely state explicitly what they want for the company from its quarters, other than expressing concern for a good image on the one hand and trying to keep costs down on the other.	Company executive in all departments, including real estate and facilities management, are equally involved in and committed to the business mission of the organization.
The facilities management and space planning staff who plan and make decisions about space are rarely involved with the primary mission of the company, and vice versa.	Business managers and facilities personnel work together as equals on building-related issues.
For most corporate executives, being accommodated is a state with no clear beginning or end, to which funds are committed on an ongoing basis with no apparent return.	Facilities is a profit center for the organization, offering cost-effective facilities services at market rates which ensure productive and comfortable employees.
Its' premises are not usually considered an integral part of the success or failure of an organization.	Cost-effective, good quality premises are considered integral to the success of the organization by senior decision-makers.

The combination of increasing business pressure towards more cost-effective expenditures on accommodation and changes in the work-force and the nature of work, as well as other societal trends, indicates that companies are following the Bank of Boston's example. More and more organizations feel that the space that an organization occupies must be an optimal quality environment that is an integral part of the performance of work. In other words, the O–A relationship is ready for a total quality approach: Businesses that embrace the concept of continuous improvement can no longer ignore the quality of their accommodation. As a recent commentator on trends in corporate real estate pointed out, "property experts still complain that users don't know what they want. Nearer the truth is the probability that users don't like what they have been given."[14]

CHANGING BUILDING DESIGN

Urban areas are cluttered with office buildings which may show some architectural variation from the outside, but which inside present standardized office interiors with fixed column spacing, ceiling light fixtures, strip windows, and services in the core. More innovative owner-occupiers are demanding something different from their office accommodation. In Europe, and to some degree in North America, a change in values and attitudes towards office accommodation is beginning to be expressed in architectural design.

At an international conference in 1992 on corporate use of architecture, the chief engineer of the Dutch insurance company Centraalbeheer described a major renovation and upgrade that had been implemented to modernize the company's headquarters outside Amsterdam. The company commissioned the building some thenty years before from leading Dutch architect Herman Hertzberger, and the building is recognized internationally as a humanistic and innovative example of corporate architecture. The challenge presented to the chief engineer was to modernize the building without sacrificing either the unique humanistic qualities of its interior space or the international recognizability of its distinctive architecture and interiors. He cited a number of key propositions that guided the company's decision-making. These include

- The office building should be used as a multi-functional instrument;
- The office building is not yet a mature phenomenon (is still evol–ving);
- The office building is the most common work environment (and should respond to people's needs);

- The office building is too expensive for a supporting role (namely, company overhead);
- The office building affects the quality of life of its users;
- Office buildings are too often ready-made and too little made-to-measure;
- In terms of space and time, the office building is used inefficiently [15]

For this company, the amount it has invested in its accommodation requires that the building actively contribute to the performance and profitability of the firm. And there is some evidence that since Hertzberger designed this highly innovative office building, several office buildings in Europe have been designed to function as more than simply office space or a symbol of the status of the organization. Most of these buildings have also resulted in distinctive architectural environments that bear little resemblance to the traditional North American squared-off, tower-like, high-rise office building. An excellent example of such a distinctive building is the headquarters of the Nederlandse Middenstands Bank (NMB) near Amsterdam in The Netherlands.

The NMB headquarters consists of ten multistory towers up to a height of seven stories, strung together in the form of an *S*, with exterior walls at an oblique angle to the ground. Designed with the explicit goal of providing a humane and comfortable work environment for the bank's employees, the building and its interiors are unique in many ways.[16] The bank's directors invoked the principles of organic architecture, including rounded forms, natural materials, and an abundance of plants and water. In spite of its large size (it accommodates some 2,000 people), the building's variegated shape was created to reduce the scale of the interior spaces and provide employees with the experience of working in smaller, more intimate groupings. The building was also designed to use solar energy and is provided with an energy-efficient heat recovery system. The tower atria accommodate abundant mature plantings, and fountains and other water elements decorate the interior public areas. The office space was designed to provide plenty of natural light and access to windows for all workers, while protecting them from glare and heat gain from the sun. Interior finishes are earth-toned wood, stone and tile, as shown in Figure 1.2, and the oblique angles of the interior walls give each workspace an intimacy and charm rarely associated with conventional office space.

In another European example, the sales team for Digital Equipment Corporation (DEC) in Finland was given the chance to replan their own office space. In contrast to the conventional process used at DEC,

Figure 1.2. Interior street and stair at the NMB headquarters in South Amsterdam, The Netherlands. Source: NMB Bank, Amsterdam.

where interior designers and space planners respond to a team manager's request for space reductions or expansion by drawing up new plans with minimal consultation with employees, the sales staff in the Finland office took over the redesign process and communicated to the design team how they felt their workspace could best meet their needs. The result is a highly unconventional office environment in which individual workstations are reduced to their minimum dimensions, employees have moveable file cabinets, and shared team spaces are designed for group work sessions, visits by clients, and employee relaxation (see Figure 1.3). All work materials are locked away at night, and employees sit wherever there is space when they next come into the office. A kitchen-lounge area provides opportunities for relaxation, and an elegantly-appointed conference room is available for meeting clients. The teamwork area is furnished with porch and garden furniture, including a porch swing, that enables employees to meet and talk in small or large groups, and which give an atmosphere of comfort, efficiency and relaxation to their work environment.[17] This workspace was no more

Figure 1.3. Team meeting space at the DEC sales office in Helsinki, Finland.

expensive to construct than a conventional office environment. The team manager realized the space reduction he had been seeking, while the sales revenues from this group exceeded those in all of DEC's other sales offices in Europe.

A recent North American example of the new kind of architecture is Steelcase's Corporate Development Center, which is in the shape of a pyramid. The social and psychological principles applied to the Steelcase building resulted in multiple work areas offering private, shared, and public space opportunities, a centrally located *Director's cluster* of managers' offices, creative and social opportunities for employees in cafeteria and break areas, and visually accessible escalators to transfer people between floors.[18] Steelcase retained experts in human communication, organizational development, and corporate facilities management to consult with the architects on this design. The building was designed to serve the goals of the organization by speeding up the new product development cycle, including new idea generation, prototype design, product testing, and marketing. The interior spaces are designed and planned to reduce isolation of individual departments and encourage them to communicate, to encourage individuals to define territorial boundaries that go beyond their individual workstation to include public and social areas, and to explicitly define processes for planning, design, and management of the workplace.

The importance of listening to employees as well as to management and the critical role of management in strategic decision-making at the level of building design are themes frequently emphasized by corporate design consultants. As his quote at the beginning of this chapter indicates, Lawrence promotes design innovations, such as those described above, as being more than the result of overactive architectural imaginations.[19] He and others like him see the emergence of a new trend in the relationship between organizations and their buildings. The keystones of this trend are as follows:

- an organization's building delivers a quality work environment to the organization at a reasonable cost,
- a building's interior environment is designed to serve a measurable purpose that advances the organization towards its performance goals,
- the work environment is a tool that, like electronic and other office technology, can increase employees' ability to get work done,
- the building provides a humane and functional work environment,
- the building is environmentally responsible,
- organizations occupying buildings no longer have to accept a compromise O–A relationship, but are increasingly seeking a mutually beneficial and enhancing marriage.

ECONOMIC AND SOCIAL PRESSURES
ON THE O–A RELATIONSHIP

Design trends like those discussed indicate an active corporate response
to the increasing costs of office accommodation, and a desire to derive
more measurable benefits from real estate expenditures. As the story at
the beginning of this chapter indicates, they also indicate a growing corporate awareness of the degree of emotional and interpersonal conflict
that issues of space and accommodation arouse. Everyone knows that
people get very emotional about space, but typical ways of managing
the O–A relationship do not address this reaction at any level. The
emotional content of the O–A relationship is an indicator that accommodation decisions are not cost issues alone. What are some of the current trends and pressures causing companies to examine their O–A
relationship more closely? Reviewed briefly below, pressures on corporate managers regarding the O–A relationship are analyzed in detail in
the next chapter.

First among these pressures are cost considerations. As companies
streamline and downsize, the costs of their accommodation become increasingly burdensome. Whether they lease or own their office space,
companies often pay heavy inner-city real estate taxes on space; further,
the office space is increasingly expensive to furnish and equip, the
amount of office technology that needs to be acquired is steadily increasing, and workers themselves appear to expect ever higher levels of
building-related comfort and services. Occupancy costs are rising for
the following reasons:

1. The increasing sophistication of modern building technology makes
   buildings a more costly capital investment, as well as more expensive
   to operate.

2. Most office employees today have at least one and often two computers per desk, along with related equipment and support technology.
   Consequently it costs more and more to equip office employees to do
   their work; it costs more because they and their equipment are occupying more space, *and* it costs more to operate that space (more ventilation, special lighting, etc.).

3. Employees expect and sometimes require an increasing number of
   building amenities, including cafeterias and restaurants, fitness
   rooms, childcare centers, and adequate parking. These amenities add
   to construction costs and are costly to operate and maintain.

4. Increasingly sophisticated building operations and office technology

require a more highly skilled, trained, and paid facilities management team to operate the building, whether the team is out-sourced to contract workers, or an in-house operation.

5. Modern office workers who are increasingly aware of the impact of the work environment on their comfort and health are demanding higher standards of building system performance, ergonomic furniture, regular environmental testing, and other costly preventive procedures.

Companies also respond to higher space costs by exploring nonoffice alternatives to reduce occupancy costs, especially for employees who do not need to be in a building on a full-time basis. These alternatives might include sales people who are on the road, data processing clerks who can work at home and professionals, such as accountants and architects, who often work in their clients' and other professionals' offices. Other companies lease inexpensive small offices in the suburbs for sales personnel who only occasionally visit their downtown home office location. Increasingly, employers favor doubling up or sharing workstations, especially for technical employees such as engineers who work intermittently on electronic workstations not dedicated to one individual. Some companies are increasing centralization to reduce their costs: groups formerly working in dispersed offices are all accommodated under one roof without reducing the efficiency of their operations. And others are exploring work-at-home programs, telecommuting, and mobile offices as options which allow employees to work anywhere, without being tied to a physical location. In fact, a newspaper report recently claimed that office buildings as a life form will soon be extinct as more of these options are adopted to enable companies to reduce the costs of accommodating their employees in the expensive downtowns of our large cities.[20]

Implementing such alternatives, however, can be slow and painful, in part because of the force of habit and tradition, but also because of the emotional meaning of space in people's lives. In response to a request for volunteers for a new work-at-home telecommuting option, one company told the program planners that working at home was impossible because of employees' needs, variously, "to be able to see each other,""to have separation and privacy," "to be accessible on the phone," "to pin up pictures," "to avoid constantly answering the telephone," "to be on 24-hour call," "to avoid being on 24-hour call," "to be seen to be part of the team," "to be seen by supervisors," "to be seen." Each opinion was advanced with the same calm assurance by people who had no intention of countenancing change but who had to conceal their intentions of blocking it.

Typically, people attach emotionally to their space at work in the following ways:

1. *Territoriality*, expressed through closure, personalization and labeling, and "adoption" of old conference room chairs and other, distinctive, items in the individual workspace;

2. *Home away from home*, expressed by living room lamps on desks, pictures and photographs on walls and partitions, sofas and easy chairs in the office, and even decorations hanging from the ceiling;

3. *Conflict*, expressed through boundary definition and defense, such as pushing movable partitions outwards so that the circulation paths get narrower, and making autocratic decisions about space use because people cannot negotiate rationally on the subject;

4. *Size and status* of offices, expressed through corner locations, and extra space and furniture for real and imagined meetings with staff and visitors.

And it is not only lower level employees who resist space change. It is often easier for managers to talk about empowerment and participatory decision making than to reduce their own office size or move away from a window. In one case where a floor had recently been completely replanned to accommodate people more efficiently, a manager in a large corner office demanded that his part of the floor be done over because of insufficient space *outside* his office for visitors! He had his own staff's circulation space and his secretary's workspace reduced in order to have a more grandiose reception area for people who came to see him.

Given the entrenched nature of many people's attitudes towards space, especially in the context of a corporate system that has used space as a reward and as a symbol of advancement, efforts to reduce occupancy costs by accommodating people in less space are likely to be ineffective if the emotional and interpersonal components of space planning are not considered. In spite of advice to the contrary, many businesses are cutting space costs by eliminating square feet—much as they are cutting personnel costs by down-sizing employees—without regard to the functions and qualities of what they are eliminating. Necessary as these cuts are, they are likely to be more effective over the long term if they are accompanied by a deeper and more sustained understanding of the O–A relationship.

Economic and social realities make it increasingly necessary for businesses to rationalize their occupancy of space. Such rationalization may result from a better understanding of how workspace contributes to employees' productivity, from a more critical analysis of the facilities management role, and from a redefinition of the work-force and the nature of office work. In reviving elements of Barnard's organizational model, and of the business-based process of allocating resources described by Bower, such analysis and understanding will ease a return to a business-driven model of accommodation strategy—not to plan plant expansion for manufacturing processes, but to determine optimal accommodation options for a given business strategy. In order to take back space-related decision making, business managers in modern companies need a better understanding of *how occupancy works*. Information on what employees need from their accommodation in order to work effectively, and on the likely impacts of various different accommodation strategies on unit productivity, will help define appropriate expenditures on space-related services in terms of value creation.

WORKSPACE: Asset or Liability

Each element of the story of the Boston property management firm described at the beginning of this chapter—the quality of the work environment, employee concerns about health and comfort, the cost of office accommodation, and the demands on today's facilities manager—are equally important reasons why the workplace is ever more significant in today's business world. Neither "space as symbol" nor "space as cost" answer the complex and interrelated questions of effective work performance, employee health, environmental quality, cost-effective expenditures, and expanding facilities management responsibilities that face most businesses of any size regarding their space.

Figure 1.4 compares two models of making accommodation decisions: the existing *cost* model that leads to less than optimal O–A relations, and the new *investment* model, in which an organization's accommodation strategy is the result of a multifactor integrated decision-making process. In the cost model, an organization's productivity is based on a centralized set of operations into which support services such as financial and administrative services, personnel services, and space planning have fed. Decisions in each of these support areas is primarily controlled by the managers of those subsections of the organization. In the cost model today, space is increasingly a liability: demand for it is going up, the costs of acquiring it and operating it are going up, pressure to improve workspace quality and comfort is increasing, and in companies that are downsizing, real estate is an easy and obvious target. Getting rid of some people and squeezing the others into less space, cutting back or outsourcing facilities staff, and clos-

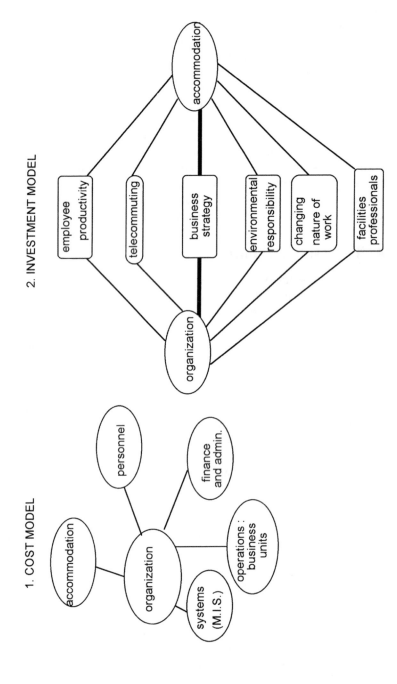

Figure 1.4. Comparison of the cost and the investment models of accommodation planning.

ing out leases on rented space all show up quickly in an improved bottom line.

In the investment model, information about a range of space-related activities is used to produce the best possible accommodation for the organization's needs. The input includes understanding how employees use space to get their work done, assessing decentralized workspace alternatives, being environmentally responsible, using building professionals strategically, and anticipating the impact of a changing labor force. In the investment model, optimum accommodation for an organization is the result of a strategic planning process whereby the production of the company's goods and services is measurably enhanced by the physical environment in which it carries out its tasks. Expenditures on accommodation are seen as an investment, the return on which is realizable in terms of greater effectiveness of work processes, improved productivity of individual employees, and accrued value of good quality and well-managed real estate. The contents of each element of the model is examined in detail in the next chapter.

In summary, the concept of a marriage between an organization and its accommodation to the mutual benefit and advancement of both sides is a more effective way to view building accommodation than as a costly and/or symbolic backdrop to the performance of work. If the O–A relationship works well, its accommodation can actually make money for an organization; if it works badly, accommodation is a costly overhead item. In the next chapter, a closer look is offered at current trends in facilities management, office building design, alternative work environments, and the impact of total quality management on the O–A relationship. This enables a more precise understanding of the critical relationship between the employees accommodated in a building and the operation, maintenance, interior planning, and management of that building.

Notes and References

1. *The Boston Globe*, 4 February 1993, p. 45. Kindelberger, R. "Law Firm Blames Building For Ills".
2. A recent publication lists the following: "facility consultants, space planners, architects, real estate brokers, engineers, interior designers, contractors, building owners, environmental specialists, city inspectors, movers, electricians, plumbers, carpenters, painters, etc." Molloy, Laurence B., "Pre-Occupancy Evaluation In Facilities Management" in *Building Evaluation* ed. W.F.E.Preiser, New York: Plenum Press, 1989 P. 59.
3. Grandiose architectural feats that have turned out to be downright troublesome to both occupants and operators include the Great Arch outside Paris, the INTELSAT headquarters in Washington D.C., and the Bateson Building in Sacramento, California.

4. Ranko Bon, *Building As An Economic Process* Englewood-Cliffs, N.J.: Prentice-Hall, 1989.
5. Figures for the 1980s showed average churn in the US to be between 27 percent and 32 percent. (Tim Springer, personal communication).
6. A growing number of companies engage in "benchmarking" their building operating costs, that is to say, exchanging detailed cost information and comparing among themselves.
7. "Planning for capital spending is a process which begins with the operating managers of a business. They are the ones who define the needs of their part of the corporation, who make the sales forecasts which justify new capacity, who review new technology to determine what the appropriate design should be, who evaluate the economics of a strategy and draft requests for capital funds and, finally, who supervise the design and construction or purchase of a new plant facility and its equipment," Joseph L. Bower, *Managing the Resource Allocation Process* (Boston: Harvard Business School Press, 1986), p. 11.
8. Chester I. Barnard, *The Functions of the Executive* (Cambridge: Harvard University Press, 1964), p. 66.
9. Barnard, *Functions*, p. 66.
10. Barnard, *Functions*, p. 67.
11. Bower, *Managing the Resource*, p. 11.
12. David K. Carr et al., *Break Point—Business Process Redesign* (Arlington, VA: Coopers and Lybrand, 1992).
13. Holly Sraeel, "Bank of Boston's JIT Gives Eileen Harvard the FM Edge," *Facilities Design and Management*, October 1992, p. 46.
14. Frank, Duffy, *DEGW News*, 2 June 1993.
15. W.M.Jansen "The Office Building as an Instrument," paper presented at *International Symposium on Corporate Space and Architecture*, Paris: 1992.
16. See Jacqueline C. Vischer and W.C. Mees,"Organic Design in The Netherlands: Case Study of an Innovative Office Building," in *Design Intervention: Toward A More Humane Architecture*, eds. W.Preiser, J.C.Vischer, E.T.White, (New York: Van Nostrand Reinhold, 1991) p. 285.
17. Frances Busse Fowler, personal communication.
18. Frank Becker, *The Total Workplace: Facilities Management and the Elastic Organization* New York: Van Nostrand Reinhold, 1990
19. Peter Lawrence, "The Design Asset," *The Corporate Board*, July-August, 1988, p. 17.
20. *The Vancouver Sun*, 2 November 1992, p. A8

CHAPTER 2

MANAGING ENVIRONMENTAL QUALITY: CURRENT TRENDS IN OFFICE OCCUPANCY

"A good fit between the building and the occupants ... cannot be accomplished by standardization, but requires a range of options within which the facility provider can offer the client some flexibility."

The Workplace Network

Of the many important changes and developments in the business world today, three significantly affect the organization–accommodation relationship. The first of these is the new profession of "Facilities Management" (FM) which has grown rapidly in recent years, to the point where there are now university-based training programs for undergraduates as well as mid-career professionals. The second factor derives from changing trends in the nature of office work. The activities that take place inside modern office buildings have changed dramatically with the advent of computers, resulting in the redefinition of roles and tasks, in the need for new approaches to the management–employee relationship, and in the definition of new ways of working. Third is the impact of electronic communications technology on worktime and workspace: with people free to perform office work from almost anywhere at any time, what is the continuing relevance and usefulness of the office building as a place to work? No longer needing to be enclosed in either space or time, the performance of work and how it is measured need to be reexamined and redefined.

Systematically examining and exploring these trends, and where they are taking modern business strategy, demonstrates the overriding importance to business managers of increasing their understanding of the critical interaction between people and their accommodation. One key to effective accommodation strategy in business is better control and management of the way workers use the environment to perform their work, leading to work environments better tailored to people's work requirements. Feedback from workers about their physical environment and how they use it is a necessary tool to improve managers' understanding of how accommodation affects performance.

WHAT DOES FACILITIES MANAGEMENT (FM) MEAN TODAY?

In a recent seminar organized by the University of Montreal's College of Environmental Design, the keynote speaker was asked to comment on the status of the architectural profession. The presentation had focused on research and practice in environmental design, and had raised issues of relevance, validity, political dilemmas, and research financing.[1] With many architects and architecture students present, the speaker reluctantly commented: "The profession of architecture seems intent on narrowing the focus of its activities, reducing its realm of effectiveness to form-giving and design, and retreating from the larger social issues which preoccupied the profession 10 or 15 years ago." There were some nods and murmurs from the crowd listening. She went on, "I would contrast this with the relatively new profession of Facilities Management, which is still at the stage of defining itself and, in so doing is embracing as many areas of activity and expertise as it sensibly can, on the grounds that operating complex modern buildings requires a professional level of competence, skill and responsibility in a diverse and broadly-defined number of areas."

This quote, contrasting the old and traditional profession of architecture with the new and modern profession of facilities management, echoes much of the current literature on facilities management today in its characterization of FM as new, growing, and hungry to embrace skills and knowledge from a range of disciplines, new and old. The president of the Canadian School of Management has described Facility Management as "an emerging and fast-growing multi-disciplinary area for career development of managerial generalists. To meet the challenges of change, it is important for managers of organizations to combine the most current technical knowledge with the ability to provide humane and effective work environments". As preparation, he recommends a training program of management education, planning, project management, financial management, real estate, and interior space planning.

Articulating the organizational goals of an FM team in his book on the organizational use of space, Steele provides a more global definition: "The overall goals for a facilities management process should be to promote a good match between users' needs and their facilities, to do this in an economically efficient manner, and to strive to create an environment that is alive and stimulating, not deadening and degrading." He adds, "productivity, cost and climate are all important variables and ... no one of them should drive out the other two because it is easier to measure or identify."[3]

In addition to emphasizing the generalist nature of this growing field, these definitions indicate the wide range of skills and activities that characterize the exercise of the profession in corporations and in the public sector. The impact of facilities management on the O–A relationship is further illustrated by the proceedings of a conference on the "Workplace Environment" in which both European and North American representatives participated. Participants defined the following important trends affecting how corporations and governments manage their space:

> Power is shifting ... client organizations are becoming more active ...occupants, rather than providers, are taking the key role [in negotiations]; customization can extend not only to buildings and services, but also to individual work settings; and individual control results in better fit and productivity.[4]

All the participants agreed that, as a result of these and other changes, the field of FM is growing in importance. Becker identifies five factors which are stimulating the growth of facilities management:

- information technology
- cost of mistakes
- global competition
- high cost of space
- employee expectations.[5]

In Becker's view, these factors cause modern facilities managers to be seen as a vital service group by their "customers"—the employees of the organization: they have moved "from the basement to the boardroom." For the new breed of facilities manager, the mandate is not just to keep the building smoothly functioning so that employees can get their work done; it is to take a holistic approach. Such an approach increases the profitability of the organization, makes it more productive and more successful, and integrates into its operations the tools employees need for their work—computer technology, office equipment, places to meet, work, and talk, telecommunications, environmental comfort, furniture, and space to relax.

In exploring the traditional separation of those responsible for facilities from the business decision-makers of organizations, Brill notes that commonly held beliefs include:

- the facility doesn't have much impact on people's performance or organizational effectiveness,
- the facility is only a piece of real estate,
- the facility is a cost center, not a benefit producer.[6]

He goes on to say that, because "none of these is true," facilities managers must start changing both other people's and their own beliefs regarding their role in the organization. Strategies for change that he outlines include increased communication with occupants and other management staff; disseminating information about the building, the furniture, and other facilities' responsibilities throughout the organization; teaching occupants to manage space and furniture better in their own workspaces; and becoming more proactive. And a recent issue of *Buildings* magazine identifies

- asset management
- space planning and large-scale refurbishment of space
- Total Quality Management programs and "benchmarking"
- streamlining space to meet down-sizing criteria
- ergonomics and human factors
- cable management
- indoor air quality, lighting retrofit and energy management

as a few of the key items that concern today's facilities managers and about which they should be informed, if not expert in the field.[7]

As the profession of FM develops in these ways, two issues arise. One is that in many large organizations, those currently in FM roles are technically-trained, "nuts and bolts" individuals without skills or interest in being in the boardroom. The kind of role expansion Becker envisions applies to a new breed of younger, better-trained manager (rather than technician) for whom a stint in real estate is a step up the corporate ladder rather than the result of a career interest in the O–A relationship.

The second, and a related issue, is the risk that in aggressively expanding the boundaries of their role and function in organizations, facilities managers are increasingly removing accommodation issues from the control of business managers and thus out of the business process of the organization. This separation of real estate management from business operations finds an echo in corporate accounting practices that show buildings and real estate as an asset on which no revenues are being realized, and the costs of operating buildings as straight overhead. As a result, facilities managers are forced to justify their budgets on the basis of avoiding costly disasters such as equipment breaking down or sick building syndrome, rather than on the basis of investment in sound facilities decisions such as preventive maintenance, efficient space standards, energy-conserving technology, and computerized inventory management. Most companies' accounting practices fail to take into consideration the value to the organization of providing a high-quality work environment to employees. The separation of facilities from the "Primary Process" of the organization means that improved employee performance, higher levels of group productivity, and reduced absenteeism and sick leaves are not factored into the justification for expenditures on buildings.[8]

In an effort to overcome these limitations and make FM more relevant to business, Binder exhorts facilities managers to alter their language and their use of terms to reflect the bottom line preoccupations of senior executives:

> We call ourselves facility managers, project managers, real estate managers, project directors, et cetera. Our titles are of minor consequence compared to our common goals as managers of corporate assets.[9]

He defines *assets* as "all the entries on a balance sheet showing the entire property of a business as cash, inventory, equipment, and real estate." Binder encourages facilities managers to convert transactions into dollars and cents in order to give senior management a more accurate idea of the assets facility managers are managing. For example, instead of stating "We bought 500 desks," facility managers should describe themselves as "responsible for purchasing $1.25 million worth of furniture"; and instead of stating "We moved 5000 boxes", facility managers have "managed the timely relocation of 1000 employees to minimize lost efficiency of the $50 million corporate payroll."[10]

In exhorting facilities management personnel to expand their roles, to take on more—and more serious—corporate responsibilities, and to concern themselves with human relations and the "people" aspects of accommodation, these commentators are encouraging facilities personnel to expand on their traditional roles as technicians, janitors, building engineers, and tradesmen and to become their organization's experts on the O–A relationship. They risk, perhaps, understating the comparable importance of providing business managers as well with some O–A expertise and a greater awareness of the impact of accommodation on performance.

INTEGRATING FM WITH BUSINESS STRATEGY

It was noted in Chapter 1 that among the reasons for increasing occupancy costs are not only the increasing technical complexity of modern buildings, but also the fact that they require increasingly highly-trained and specialized staff. As a result, rather than pay ever larger salaries for building operation and maintenance, many companies are out-sourcing FM services, that is, buying the services from outside purveyors, or they are structuring and preparing their facilities management team to compete on the open market in providing services to other companies.

Whether in-house or out-sourced, and regardless of corporate size, the financing of facilities management services is changing, and these changes directly affect the relationship between real estate and business units in an organization. A team from MIT spells out five coping strategies in real estate financing that reflect stages in the maturation of the relationship between a firm and its FM department.[11] At the first stage, buildings and their accouterments are supplied to business units on a demand basis, regardless of cost. The "wants and needs of the business units drive the process." The second stage occurs once "real estate has come to the forefront of management's attention" and costs are cut back. This stage has the real estate executives cutting back on goods and services supplied to business units through the invocation of space and furniture standards and the application of fiscal control mechanisms and other cost-cutting measures. By the third stage, the real estate department has adopted market standards for cost and quality and is beginning to develop a revenue-accountable relationship with the business units, who in turn are beginning to take some financial responsibility for their accommodation needs. At the fourth stage, real estate services compete with the market in terms of rents, costs, and quality of services and space provided: "Since the product is built to market standards and earns a market rent, the real estate is no longer a source of subsidy to the business unit from the corporation." Some real estate groups at this stage increase their profit by offering their services to other corporations. By the fifth stage, the real estate process is "advancing the business mission" by acting as a "strategic unit whose charge is to use real estate decisions to further the mission of the firm." Real estate is informed about the business units' business plans and ensures that they incorporate a real estate strategy. In having to make a business case for their accommodation planning, business units' real estate decisions become in fact strategic business decisions.

The five stages are summarized in the table, below.

Table 2.1.   Five Stages of Real Estate Financial Coping Strategies

Real Estate Coping Strategies	Impact on Facilities Managers
1. Engineering approach	Managers are service providers, making decisions based on budgets received from the organization
2. Cost minimization	Managers try to cut costs so as to have smaller budgets, while clients' demands for quality of services increase.
3. Market cost and usage standards	Managers start to see themselves as eventually competing with the market, so begin to develop standards and rationalize budgets.
4. Market design approach	Managers receive budgets from "clients" in business units, so cost and quality of services are tailored according to market conditions.
5. Business strategy approach	FM competes directly with outside vendors, but offers better services, being positioned inside the organization. FM is part of business strategy of the organization.

This analysis shows how building concerns can become progressively integrated into business strategy, indicating how economic and financial considerations can have the effect of shifting corporate attitudes towards real estate over time. This progression appears inevitable in view of the pressures on FM to change in the current business climate. As Handy points out in his book on learning and change, not learning in time to change means the eventual disappearance of the [FM] organization.[12]

THE IMPACT ON SPACE USE OF CHANGING TRENDS IN THE NATURE OF WORK

One of the legacies of the corporate separation of FM from business units has been the concentration of facilities managers on the technical aspects of building operation, at the cost of a better understanding of "the human aspect." There is an entire realm of knowledge and understanding that pertains to the human use of space that falls through the cracks, as business managers manage the performance of their workgroups and facilities managers manage the physical environment of employees. As a result, no one officially takes responsibility for the interaction between worker behavior and their workspace environment and how this interaction affects the performance of work.

Several of the pressures identified in the previous chapter indicate that managers continue to ignore this relationship at their peril. Changes in the demographics of the work-force, changes in the way office work is performed, changes in employee awareness and attitudes towards occupational health, and new possibilities in the location and timing of office work are all critical factors, directing future-oriented businesses towards a closer and more systematic examination of the worker-environment interface. Just as business managers need to understand competitors' behavior to formulate their own business strategy, both business and facilities managers need to understand the user-environment interface—or system—to formulate accommodation strategy for business units as exemplified in Stages 4 and 5 of the 5-stage model.

The use of workspace in office buildings is being affected by some important changes in the organization of office work. Of the critical factors currently influencing the way workers use space in offices, three to be discussed here are

- the movement towards "knowledge work" and group and individual creativity,
- changes in clerical and support work, and
- the availability of alternative work environments.

The critical changes in employee awareness and attitudes towards occupational health and the perceived financial burden on companies to provide more, and more human, services in their buildings will be dealt with in depth in the next chapter.

Team-work and Creativity

As routine office tasks are taken over by technology, companies rely as much on the knowledge, competence, and creativity of their employees as on their ability to complete repetitive assignments within a given timeframe. There is a growing proportion of "knowledge workers" in most companies, and the traditional distinction between management and clerical work is breaking down as more technical and professional people move into corporate life.

The value of knowledge workers to their employers is in their creativity, in their problem-solving, and in their innovative ideas, not in executing paper-based tasks in standardized environmental modules. As a result, their workspace must accommodate nontraditional office behaviors, teamwork, informal group meetings, odd hours, a range of supports, and shared project space. The space they work in can clearly mitigate *against* the creativity of such people, by inappropriate, uncomfortable, and dysfunctional space design, that inhibits communication and reduces the performance of knowledge workers for their company. Traditional office layouts, with enclosed individual workstations along corridors, a few large shared conference rooms, and enclosed offices for managers are not as useful in supporting creativity as an office environment that encourages teamwork, facilitates project work, and allows easy communication between team members and team leaders—and yet enables individuals to be private when they need to concentrate. As long ago as 1985, a psychologist and a architect published their "cave and commons" theory of workspace lay-out.[14]

Certain buildings designed with "teaming" in mind (such as some of the examples described in Chapter 1) have incorporated physical elements that encourage an *esprit de corps*. These include the clustering of work-group areas along an interior "street" off which shared or centralized facilities such as meeting rooms and coffee lounges are located—such as the NMB headquarters—and the porch furniture used by DEC salespersons in Finland. The Swedish national telephone company has actually grouped its workers into neighborhoods, complete with interior village greens, fences, and shops to mark off team boundaries.[13] These office designs aim to facilitate comminication and encourage creativity without eliminating individual privacy and a sense of belonging.

Changes in Clerical and Support Work

The secretary role has been dramatically redefined in modern organizations. Not only are men entering what has usually been considered a woman's job in ever greater numbers, but the work itself has expanded to be more than typing, bookkeeping, and telephone answering. The office technology revolution has allowed most of these functions to be managed electronically, and this, coupled with technical and professional people working in groups, has enabled the one-time secretary to take on an expanded administrative role. As many teams and work-groups function today without a clear hierarchy, the person often hired as secretary to a manager becomes the de facto administrative officer for the whole group. He or she handles human resource and budgetary management, controls information flow and communication, and takes a strong role in planning and deployment of the group's resources.

Despite these changes, lingering traditional definitions of the job of secretary mean that spatially this person is seldom adequately accommodated, having to carry out responsible and often confidential administrative duties at a desk or cubicle adjacent to a manager's office. As a result, it is not uncommon for people traditionally designated as "secretary" or "administrative assistant" to complain bitterly to whomever will listen that noise interrupts their work, that passersby and visitors look at their screens or desks when they are working on personnel files and confidential material, and that there is not enough space in their cubicle to store all the files and documents needed by their group. This feedback can be dismissed by managers, or it can be incorporated into new space decisions that recognize the changed definition of the secretarial role.

Clerical work itself has become vastly redefined. People who pushed papers, filed and typed, and copied figures from one book to another in the traditional office have been replaced by data entry, data processing, and other computer-based functions. Evergrowing numbers of office workers now sit at terminals for most of their working day and are often the least comfortable group in the building. A shift towards the proliferation of service organizations in our economy means that the clerical staff corporations employ to interface with their customers work all day on the telephone and at a computer screen, often with a second screen or electronic reader on their work-surface.[15]

These employees are reservations clerks for transportation companies, such as airlines; they are customer service clerks for catalogue companies; they are operators for the telephone company; they are accounts receivable clerks for any company that serves the public and bills for its services.[16] In many cases, these workers, who perform only a single series of repetitive tasks, are neither provided with enclosure nor with any office amenities. They sit close together, have little if any desk storage, and occupy a standardized basic workspace that they are not allowed to move away from. They are generally planned into a hierarchical layout, with a supervisor sitting where all the employees of a work-group can be seen. They are programmed into a rigid schedule of coffee and lunch breaks because the telephones and terminals must always be staffed. They often suffer from neck and shoulder pain, carpal tunnel syndrome, and eyestrain—all of which could be alleviated by better environmental planning and task management.

Of all the types of office work that could be enhanced by carefully designed, ergonomically responsible work environments, the performance of these "workers on the electronic plantation" stands to provide the most measurable benefit to employers. Yet environmentally these groups are often the worst served in an office building. The density of their equipment raises the temperature of their work area and dries out the air; the long hours at a screen require special lighting that they do not often get; any need for personal space or privacy is overridden in a typical layout that gives priority to the need to see and be seen by coworkers and supervisors. Although these are not highly paid workers, they can have substantial responsibility for the basic cash flow of the company—they are often "its bread and butter." Investments in improving their work environment should equal or at least parallel investments routinely made to accommodate computer equipment correctly: clean air, stable temperatures and humidity, and protection from static electricity and vibration.

Changes in the Space and Time Constraints on Office Work

As companies continue to out-source more functions and services, there will be fewer permanent employees at any level.[17] Eventually, two classes of employee will emerge: the temporary, lower paid, contract staff who can be laid off when their skills are not needed, and the permanent, secure and valued employees who will receive all the benefits of long-term employment.[18] The technology to enable staff to work at home, at hotels, in airports and in other people's offices as well as in their home office is now available, and companies are trying to reduce the amount of permanent space staff occupy. Some companies, like IBM, have already made dramatic real estate reductions. Such changes will increase as numbers of temporary staff grow, because they will work for more than one company at a time, and will not need traditional commercial office space in order to do what they do. Strong advocates of "tele-commuting," however, while endorsing video-conferencing and picture-phone technology, have not yet found a substitute for the value of face-to-face communication for certain types of work.

Companies are already examining alternative workspace opportunities that free their employees from long hours of commuting, allow them to work in clients' offices, and take advantage of up-to-date communications technology. As workers move into telecommuting mode, so space in the main office is replanned to increase efficiency and take advantage of part-time attendance. Some of the alternatives entering the vocabulary of space planners and corporate executives are:

- telecommuting
  —the virtual office
  —work-at-home offices
- satellite offices
- the nonterritorial office
  —just-in-time (JIT) offices
  —free address systems
  —hotelling
- universal planning

A brief discussion of each of these alternatives follows.

Telecommuting connotes the technology that makes possible a variety of alternatives to the fixed geographical placement of the office possible. It is known as the "virtual office" when the employee is provided with the appropriate equipment to set up an "office" and work almost anywhere. This is appropriate for sales personnel, who need to be able to work and communicate with colleagues and supervisors from airports, cars, hotel rooms, and other people's offices. Telecommuting also means the possibility of working at home. So far, two types of office workers favor work-at-home offices: highly trained and highly paid professionals, such as lawyers, who take their work home and do not replace their space but rather extend their working day by continuing to work at home outside normal working hours. The other type of work-at-home initiative has been widely implemented in Britain and applies to skilled contractual workers who perform a standardized service for an organization often on a piecework basis.[20] These employees do not need to be physically accommodated in the company's office space because they interact primarily with equipment and can perform their tasks with minimum contact with other company personnel. Not more than 15 per cent of a company's personnel is generally eligible for teleworking from a home office.[21]

Satellite offices are an option for this type of worker as well as for salespeople. Unlike the "virtual office", these are physical spaces, equipped with the technology that employees need to perform their tasks. They are decentralized to respond to employees' needs for office accommodation in the vicinity of their home or their sales territory; they can be found in shopping malls and other suburban sites. They are an alternative for workers that do not need to be "downtown" but do not want or need to work at home. The satellite office may be provided by a company for its own telecommuting staff (for instance, Nynex in Massachusetts), or may be available for public use on a fee-paying basis. Of necessity, the satellite office provides a standardized work-place module that can be used interchangeably by different users, and in this respect it represents one interpretation of the nonterritorial office.

The *non-territorial office* is a general term used to cover a wide variety of standardized work-spaces that are used interchangeably by company employees or contract staff. IBM Canada, Arthur Andersen Consulting, and some large accounting firms have experimented with a system they call "hotelling", whereby standardized, well-appointed, enclosed offices are available for use by professional staff who are only intermittently in the office, on a reservations-only basis. The small number available ensures that they are not vacant. Since professionals, such as accountants, are expected to work in their clients' offices, they are not often in their own office if they are good at their jobs. Other types of hotelling involve a more modest, standardized cubicle that can be used by sales staff or temporary personnel on a first-come, first-served basis.

Another expression of workspace interchangeability is the *Just-In-Time* concept (JIT), developed in Japan as a way to speed up assemblyline activities in manufacturing environments. It has been tentatively applied to certain types of standardized white-collar tasks that involve the rapid and efficient processing of paper. The JIT concept means that each team member is equally capable of carrying out the tasks of other team members and can therefore be active at any stage of the process that needs resources. The work environment is designed to be flexible in response to work-flow analyses; it is standardized to ensure interchangeability of people and tasks, with movable chairs and tables, movable cabling, and a clear and unobstructed visual field. A good example in North America is the Bank of Boston's office building in Canton, Massachusetts, described in the Chapter 1, which recuperated its investment in reduced space costs after one year.

The *free address system* takes the interchangeability of workspace one step further by providing a completely open-plan standardized environment to technically trained staff who work in teams on projects and need above all to interact and communicate. In these offices, people sit at tables rather than desks, with no permanently assigned positions. They move their files around with them in individual mobile file cabinets. In this type of organization, therefore, people rely heavily on centralized and accessible file storage. For concentration or work requiring privacy, open-plan work areas are augmented by one or two enclosed offices that are designed to be shared. Each space serves a team for the duration of a project, and then team members disperse to reform in a different configuration. A good example of this arrangement in North America is the headquarters of the National Association of Home-Builders Research Foundation in Maryland.

Universal Planning (UP) is another way to make individual work areas interchangeable, in this case by standardizing floor layouts and configuring them with standardized office furniture. UP layouts consume less space, because they use systems furniture, and they are designed to function for a variety of work-group configurations. When work-groups change or are moved, only the people move, along with their telephone numbers, and their workspace accommodates whomever replaces them with no physical adjustments to the space, for example, in the San Ramon, California, headquarters of Pacific Telesys. Universal Planning can be applied to a variety of physical locations, from the floors of the highrise office building downtown to a number of nonterritorial and satellite office alternatives. Although the initial investment in re-planning, furniture, and technology can be significant, the payback on reduced space costs, significantly reduced churn, and overall increased flexibility is estimated by one company to be no more than 3 years.

In summary, the impact of organizational, technology, and economic changes on space planning and facilities management is extensive and long-term. Facilities departments alone cannot handle these changes; they need to be addressed by business managers as part of strategic planning, involving, as they do, changes in personnel, skills, technology, and space. Indeed, to undertake to invest in a just-in-time office design, or to send a work-group home with computers to telecommute, without linking such a far-reaching decision to staff retraining, performance measurement, social and cultural values change, investment in new technology, new management techniques, and overall corporate strategy, is to risk business failure.

MANAGING THE HUMAN ASPECT

With trends and changes like those described above taking place at an ever-increasing pace, managers responsible for accommodation planning are increasingly considering ways of informing, and being informed by, employees in their capacity as building users about their work environment. Major changes in the work environment, and the way that environment is managed, cannot be successful without employee "buy-in" to shared goals and objectives, and may benefit from employee participation in defining and designing environmental change. Managers are beginning to develop methods and techniques for communicating effectively with building users about their environment so as to better understand their environmental requirements and to engage them in implementing accommodation decisions. Communication between building managers and occupants has become increasingly important for today's facilities managers, although the mutual exchange of information runs counter to the traditional, backstage, unseen support role FM staff have tended to favor in the past. Two-way communication about the work environment is a key element in shifting from a "reactive" to a "proactive" model of FM. It is crucial, first, to the redefinition of FM as having a "human" and not just a "building" orientation; second, to the empowerment and involvement of employees in good use of their workspace; and third, to the awareness of business managers of accommodation decisions.

The table below compares the reactive with proactive ways of making FM decisions. It contrasts the traditional approach to managing space costs with a more modern and human-oriented approach to making accommodation decisions.

In their capacity as technicians, reactive building managers field requests and complaints from users and try to solve their problems and provide them with a comfortable work environment within the constraints imposed by building systems, organizational policies, and budgetary restraints. The difficulties of meeting these often conflicting sets of requirements can result in an adversarial relationship between users and managers of buildings that has a negative impact on the productivity of both sides. By improving communication between building users and building managers, both have an opportunity to recognize the organizational and physical constraints on problem solution. Users learn more about what their buildings can and cannot provide, and building managers become more actively involved in the tasks and mission of the organization.[22] Ellis points out that good space planning solutions tend to lie more in a good planning process than in seeking out

Table 2.2. Reactive and Proactive Ways of Making FM Decisions

Reactive	Proactive
The building is seen as a back-drop for getting work done. It costs the organization money to operate, so good accommodation decisions are based on effective cost-cutting.	The building is a tool for effective work performance by employees. Investing in a good work environment is seen as a way to generate a financial return to the organization.
Planners feel that there is a single best way to fit people into a given space, if only they could get it right the first time and people did not keep complaining and moving things around themselves.	There is no single best way to fit people into a space because space needs depend on the nature of work that is constantly changing. Space planning is a negotiation, and changing a configuration does not mean failure.
The corporate value known as the "currency of space" means that space and amenities in buildings are provided on the basis of rank, status and reward.	Effective space planners use space and amenities to meet employees' functional needs, and best serve the goals and purpose of the organization.
Effective building management stays "behind the scenes" keeping things running smoothly and excluding building users as much as possible from decisions that get made about the building.	Managing human relations is part of facilities management, and effective building managers incorporate communication with users systematically into the decision-making process.
An adversarial relationship between building occupants and building managers results in users making demands of managers and managers reacting as best they can within the constraints of building technology, and facilities budgets.	A cooperative relationship between building occupants and building managers enables managers to take responsibility for activating and managing communication about building-related issues.

a "best fit" physical layout. A negotiating process that leads to spatial layouts that both sides not only accept but enjoy enables managers to respond to intrinsic as well as to extrinsic pressures for change.[23]

As managers become more adept at negotiation and communication with building occupants, they are in a sense dismantling and removing rather than defining and strengthening the boundaries of their profession. By disseminating information about building use and operation, facilities managers are inviting building users to participate in building-related decision making. By encouraging feedback from building users about their work environment, building managers open up communication on building-related issues, encourage business managers to participate in accommodation decisions, and potentially expand organizational definitions of corporate accommodation to include issues that go beyond the immediate building. In companies where facilities managers are ready for the boardroom, this kind of redefinition of the FM role may seem appropriate; but in companies where FM is still only a technical and support function, this redefinition risks moving O–A relationship responsibilities out of FM hands altogether.

CURRENT TRENDS AND THE HUMAN ASPECT

Different companies will likely develop different models of FM according to their needs and corporate values, but one thing is clear: responsibility for accommodation is developing as a legitimate corporate function as companies become more cognizant of the implications of their accommodation decisions. There are certain distinct trends that characterize this development. First, the new interest in reengineering work processes to increase efficiency and optimize the deployment of personnel encourages business managers to consider workspace more seriously. Decisions to add or move people, to change office technology, or to retrain work-groups, imply greater acknowledgment of related spatial implications such as flexible workspace needs, changing heat and light requirements, and adaptable group spaces. Another important trend is the result of societal changes in the nature of work. Companies need different skills and abilities from their employees than they have in the past. They no longer need people for mindless clerical or machine-based tasks, replacing these with computers. They require employees and contractors to take more responsibility, to act more autonomously, and to adapt to an ever-increasing choice in terms of how, when, and where work is done in order to provide good quality results rapidly.

And finally, a critical trend is towards the increasing knowledge that is available to managers and space designers regarding the impact of the physical environment on people and on the performance of work. Whether this knowledge is used by a more proactive FM team that incorporates the human aspect into their space management activities, or whether it is used by business managers to make better accommodation decisions for their staff, the inclusion of the people aspects of accommodation is increasingly easy and accessible for effective accommodation planning.

As a result of these trends, feedback from building users about their use of space and their ability to adapt the work environment to meet their needs is an important step towards making their accommodation an effective tool for work. As facilities management develops in corporations, proactive accommodation decisions that attempt to incorporate the human aspect become more apparent. According to the 5-stage model of FM financing, the Market Design stage is when FM first starts to define its "client," and must therefore respond to clients' requirements; and at the Business Strategy stage, FM has to not only meet clients' needs cost-effectively, but also competitively. Therefore, at both these stages, feedback from building users regarding how people use space, what they need to work effectively, and how to make their accommodation pay off in terms of helping them get work done is likely to be a useful tool.

In the next chapter, the nature of this type of feedback and how to acquire it is explored in more detail. How do we know what questions to ask building users? How can we get useful information about their workspace without eliciting a long list of demands and complaints? How do we use the feedback once we have it, and apply it to defining an improved accommodation strategy? The next chapter focuses on attitudes of building occupants towards their physical environment, and reviews some of the arguments advanced by business strategists concerning the relationship between real estate expenditures and employee productivity. A primary topic for the next chapter is the nature, function, and definition of feedback from building occupants in order to develop a better understanding of the relationship between people's work activities and their physical workspace. Methods and techniques are explored for eliciting the right kind and amount of information from employees so that this can be usefully applied to the design of an organization's accommodation strategy.

NOTES AND REFERENCES

1. Université de Montréal, Faculté de l'Aménagement, "La recherche et les pratiques de l'aménagement," Colloque sur l'aménagement, Montréal, 14 October 1992.
2. Quoted in Yvonne Bogorva and Wojciech Nasierowski, "Different Types of Managers: Positioning the Facility Manager," *Canadian Facility Management*, March–April 1990, p. 31-35
3. Fritz Steele, *Making and Managing High Quality Workplaces: An Organizational Ecology,* New York: Teacher's College Press, 1986, p. 41.
4. "The Workplace Network: a forum for learning and sharing" Report on an international workshop, Sweden, 1991, Ottawa: Public Works Canada.
5. Franklin Becker, *The Total Workplace: Facilities Management and the Elastic Organization,* New York: Van Nostrand Reinhold, 1990 p. 9.
6. Michael Brill, "Achieve Success By Changing Attitudes," *IFMA Journal*, Winter 1988, p. 15.
7. *Buildings* 87, no. 3 March 1993.
8. B. Bleker, and L. J. Regterschot, "Facility Management In The Netherlands," *IFMA Journal*, Winter 1988, p. 24.

9. Stephen Binder, *Strategic Corporate Facilities Management*, New York: McGraw-Hill, Inc. 1992, p. 2.
10. Binder, *Strategic Management*, p. 30.
11. M. Joroff, M. Louargand, S. Lambert, and F. Becker, "Strategic Management of the Fifth Resource: Corporate Real Estate" Report of Phase One Corporate Real Estate 2000: Industrial Development Research Foundation, 1993 pp. 50–52.
12. Charles Handy, *The Age of Unreason*, Boston: Harvard Business School Press, 1990. p.23
13. I. Soderberg, "Studies of an Organizational Change: the work of telephonists and office design" Conference presentation, *Corporate Space and Architecture/Territoires et Architectures d'entreprises*, Paris, France, July 1992.
14. Philip Stone, and Robert Luchetti, "Your office Is Where You Are," *Harvard Business Review* March-April, 1985, pp. 102-117.
15. Roger Swardson, "Greetings From the Electronic Plantation," *City Pages*, 21 October 1992.
16. Called "bioware" by Dr. Tim Springer, these employees are seen as "integrated elements of a technological system" (personal communication).
17. Handy, *Age of Unreason*, p.57
18. *The Montreal Gazette* 1993 Hadekel, Peter "Very soon the working world will be divided into two types of people." Also, in Britain, legislation has been introduced to protect the health insurance and other benefits of "home-workers" whose number is growing rapidly as companies take advantage of telecommuting opportunities.
19. Lotte Bailyn, "Toward the Perfect Workplace?" *Communications of the ACM* 32 no. 4, April 1989: pp.460–470.
20. Bailyn, "Toward the Perfect Workplace" pp. 460–470.
21. "Telework 94," Conference in Toronto, Canada; 1–3 October 1994.
22. D. B. Frewald, "What we have here is a failure to communicate," *FM Journal*, May/June 1993, p. 18.
23. Peter Ellis, "Toward the Organic Office," *Facilities* 9, no. 4 October, 1991 8–12.

CHAPTER 3

Using Occupancy Feedback: A Strategy for Managing Workspace Improvements

> *"Ten years of effort have established Facilities Management as a discipline but have not overcome the inertia caused by decades of missing feedback between organizational performance and the design and management of the physical workspace."*
>
> Frank Duffy

HEALTH RISKS IN THE OFFICE

A women's magazine reports that 70 per cent of people working in office buildings in Montreal and Toronto in 1993 complain of indoor air quality problems. What this apparently means is that workers in 70 per cent of the office buildings in these two cities express, when asked, concerns about indoor air pollution. The article warns its largely young, female readers to beware of the air in sealed, high-rise office buildings, implying that their health might be at risk.[1] This type of journalism is far from unusual where indoor air quality is concerned. Hardly a magazine, newspaper, or newsletter in recent years has not carried at least one story on indoor air pollution and the implied threat to the health of office workers, to the point where office workers everywhere are concerned about the air quality in their buildings. However, these concerns are often incompletely linked to the objective facts of ventilation systems' operation. These topics are discussed in detail in Chapter 5.

This growing awareness of possible health and comfort threats in the workplace has affected employees' expectations of corporate responsibilities regarding environmental quality at work. Many leases are now only signed after assurances from landlords that ventilation standards are met in the building, and many of the newer buildings commissioned in North American cities have incorporated expensive up-to-the-minute air handling technology to ensure likely tenants and future owners that no one will be able to accuse them of polluted indoor air. The story of the Boston property management firm in Chapter 1 illustrates what a lengthy and costly mistake indoor air pollution—or even the threat of it—can become.

Along with heightened anxiety about indoor air pollution, office workers have begun to worry about eyestrain, fatigue, and headaches in the afternoons, cumulative muscle and nerve problems in wrists, necks and shoulders, as well as lower back pain. Unionized employees are represented by negotiators arguing for mandatory rest periods during the work day, better lighting, and ergonomically responsible furniture for workers performing screen-based work. Responsible employers attempt to provide safe and comfortable work environments by designing workspace to meet existing health and comfort standards. Nevertheless, at least in North America, lawsuits are proliferating and health insurance premiums are mounting as increasing numbers of work-related illnesses and injuries are reported by workers in office buildings.[2]

As well as contributing to the costs of office space by forcing employers and landlords to meet a rising number of health and comfort standards, environmental awareness and concern among office workers generates a difficult dilemma. Whereas everyone wants to reduce illness and injury, the line is often blurred between illness and injury on the one hand and discomfort and anxiety on the other. Cost conscious managers fear spending exorbitant amounts on better furniture, new lighting, and individual ventilation controls, without proof that *not* doing so will harm workers. And if *not* doing so will *not* harm workers, then managers need to know that spending money on such changes will measurably *improve* workers' performance to justify the expense.

For example, before moving into their new, purpose-built headquarters in San Ramon, California, some years ago, Pacific Bell's large systems engineering group asked for fully enclosed individual offices for each engineer. The cost implications of this change in an open plan building were such that the system engineers were offered a choice. Either you guarantee an increase in your productivity by a factor of three (be three times as productive) to generate enough revenues to warrant the extra expenditure of some $3 million on enclosing your offices; or your work performance stays the same, in which case your work performance means your enclosed offices cost too much to implement. Needless to say, the engineers still work in partitioned workstations like everyone else in the building.[3] On the other hand, a recent study of absenteeism at Apple Computer led managers to conclude that the too-open workstation layout was in fact tangibly decreasing their systems engineers' productivity, and they found it worthwhile to spend the money to replan the work environment so that more rooms and a greater variety of enclosed opportunities for concentration, work sessions, and meetings were available to these employees.[4]

Whether decisions about enclosed or open office layouts rest with facilities managers, business managers, senior executives, or a management team, most of these types of choices are not clear-cut. In fact, in spite of all the research that has been done, there is presently no rational and strategic way to decide when and how much to spend on the physical environment of work so that

- employee performance is maximized
- employee health is not threatened
- such expenditures are cost-effective
- such expenditures are a strategic investment.

THE PRODUCTIVITY DEBATE

Trying to link accommodation costs to worker productivity has been widely invoked as a rationale for office expenditures. This link was forcefully expressed in an early 1980s study of the impact of the work environment on productivity, in which physical elements such as office enclosure were measured for their contribution to employees' productivity by relating them to the salary received by each worker.[5] Studies continue to proliferate showing how environmental improvements have or have not resulted in measurable productivity increases.

The Army Corps of Engineers published the results of a study measuring the impact of ergonomically-designed systems furniture on the productivity of a work-group as compared to those in conventional furniture. The study compared a "productivity index" before and after installation of two groups of new furniture, ergonomically-designed systems and conventional, and found that whereas satisfaction improved in both groups as a result of the new furniture, productivity was only enhanced in the ergonomically-designed systems furniture group. The researchers computed the payback on systems furniture in terms of both space savings and increased productivity to be 10.8 months.[6]

In a similar study carried out at Aetna Insurance company, work output among workers with new furniture and a new layout in an otherwise unrenovated office space increased 67 per cent. When, in addition, the space was renovated, a smaller increment was noted, generating an overall increase in productivity of 53 per cent over an established and premeasured base rate. Data on absenteeism were less clear, showing a decrease of 14 per cent after the introduction of the new furniture, and a small increase of 7 per cent after the space was renovated. Comparing the costs of the new furniture and the renovation to the dollar value of the measured productivity gains, the author estimated that the value of half of the productivity increase paid for the environmental changes. Thus all the changes were paid for within 2 years with a net benefit to the company of $110,000.[7]

In an exhaustive review of studies such as these, Springer concludes that environmental improvements, combined with improved communications and electronic technology, are most effective if carefully targeted to the tasks of specific employees. He cites studies in addition to those mentioned above in which the introduction of new furniture, task-appropriate lighting, and sound-hoods to reduce the noise of impact printers have all been linked with improved worker performance. Some of these improvements have occurred within the context of moving to a new facility or a large-scale renovation, and productivity improvements have resulted from the combined impact of the new environment and the up-to-date technology that has gone along with it.[8]

Estimates of the effect the work environment on employee productivity typically range from 5 per cent to 25 per cent, using a wide range of productivity measures.[9] When employees were asked to estimate their own productivity, a recent British study reported that 70 per cent of them estimated that their productivity was adversely affected by an uncomfortable work environment.[10] If, as is estimated by MIT's Office of Facilities Management Systems, for every $7 spent on operating, equipping, and maintaining a building, another $70 is spent on staffing it—namely on the faculty members and their research and teaching assistants—then the dollar return on investing in building quality can easily be justified in terms of optimizing staff performance.[11]

Springer points out that a simple dollar-based calculation of employee productivity is usually not possible. The dollar amount a company spends on new furniture, for example, is part of its capital budgeting process and is subject to depreciation allowances, present value calculations, and the company's estimated internal rate of return. The dollars represented by an employee include salary and benefits and are based on an estimated efficiency ratio, as no one is productive 100 per cent of the time. Moreover, not all of an employee's tasks relate to one environmental element: most employees carry out a variety of tasks during the day, including meetings outside their individual workspace, VDT work, and telephone use. He concludes, "Whether using 'simple' payback, payback based on after tax cash inflow or the results of net present value calculations, ... the results of the different methods of calculating the bottom line impact suggest disparate interpretations of the magnitude attributable to the gains in productivity"[12]. What is not in question, he concludes, is that there is a measurable and quantifiable gain. He cites additional examples which show that investments in energy-efficient lighting systems, improved HVAC performance, and improved building maintenance yield savings not only in operating dollar expenditures on buildings but also in improved worker performance resulting from more task-suitable lighting and better control over temperature and ventilation conditions. Thus in one of his examples, the payback for a new energy-efficient lighting system was only 73 days.[13]

In spite of the demonstrated complexity of the environmental improvement-worker productivity relationship, business and financial managers still tend to base their decisions about what and how much to spend on the work environment on presumed or anticipated *measurable* productivity increases. Figure 3.1 below illustrates three alternative premises of the productivity debate. Either people become increasingly productive for each dollar spent on their accommodation (line A), so the more money spent, the more productive the worker. *Or* people only increase their productivity up to a certain point relative to the amount spent on their accommodation, and then level off (line B), so it is important to try to define when that optimum point is reached. *Or* workers become increasingly productive up to a certain level of expenditure on accommodation, then work performance drops off because they are "too comfortable" (line C), so it is important to stop improving their environment before they relax too much, fall asleep, get spoiled, etc.

The diagram illustrates the difficulty for managers of identifying the critical point at which improving the office environment pays off maximally in terms of increasing occupant productivity. In fact, using employee productivity measures as the basis for calculating appropriate expenditure on the office environment is a weak and ineffectual argument for them. If this argument is presented exclusively as a causal relationship—namely that increasing the quality and/or comfort of the

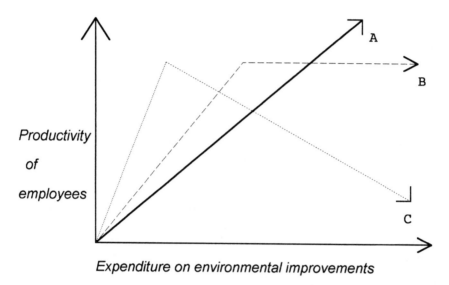

Figure 3.1. Linear relationship between quality of the work environment and productivity of employees.

work environment directly increases people's productivity—the obvious result is that fewer and fewer ever more productive workers will be needed to perform the same tasks. To reduce this argument to its absurd conclusion, in the perfect work environment, no people will be needed to do the work at all!

The productivity argument, then, although supported by research—a better environment *does* help people work better -- does not easily offer criteria for managers to decide, for example,whether lighting or ventilation should be improved, what form the environmental improvements should take, and the relative advantages of new furniture over noise reduction or better indoor air quality. From the viewpoint of facilities managers, the improved productivity argument is a hard sell to cost-conscious executives who are not yet prepared to strategize with them to define the optimal workspace for employees.

FEEDBACK FROM BUILDING OCCUPANTS

Senge points out that people use the term "feedback" to mean "to gather opinions about an act we have undertaken"[14]. But relative to a system such as the user-environment interface, feedback is a broader concept.

"It means any reciprocal flow of influence ... it is an axiom that every influence is both cause and effect. ... [As such] the feedback loop overturns deeply ingrained ideas, such as causality." He argues that feedback in this sense places the human actor *in* the system and as part of the feedback process, not standing apart from it. "This allows us to see how we are continually both influenced by and influencing our reality."[15]

In terms of this argument, positing a causal relationship between workspace and productivity is not a satisfactory basis for projecting accommodation expenditures because the worker and the work environment are an interactive system. As building managers know, a suspected air quality problem, such as an odor, or headaches and fatigue in the afternoons, whether or not a cause is found, can reduce people's effectiveness by causing them to stop work, complain to each other, complain to facilities staff, leave their desks to go outside, move into another room or floor, or even leave work early and go home. Behaviors such as these reduce human effectiveness and give the organization less than it is paying for from both the worker and the building.

In another typical example, facilities staff received persistent complaints from a woman whose work required close reading and writing of documents. She reported headaches, dizziness, and sore eyes. She thought there was an air quality problem in her office. Eventually, her physician prescribed glasses, but her complaints persisted. The problem was ultimately diagnosed as glare reflections from the glass panel that covered the surface of her desk. The panel was removed and her symptoms disappeared. This solution was an inexpensive one that had cost a certain amount of money to diagnose. Is such expenditure warranted on each and every employee on the grounds that they are likely to become more productive if such problems as these are corrected? Or is there a generic way of diagnosing more broadly problems that might be affecting groups of workers and using this information to guide priorities for environmental investment and improvement?

In fact, employee awareness of health and comfort issues such as indoor air quality can be turned to advantage if employees themselves are used to provide diagnostic information about how well their workspace functions. The knowledge employees have about their accommodation, and how they use it to get work done, can, in line with Senge's definition of system feedback,

- be used by corporate and facilities planners to generate more appropriate accommodation for an organization;

- translate into a better corporate ability to create a functional work environment that furthers the performance of work; and

- establish a basis for negotiation on environmental issues between landlords and tenants, occupants and managers, and clients and designers, regarding the ultimate quality of the work environment.

In short, feedback from workers themselves can help managers turn the work environment into a tool for work.

Some corporations and government agencies have already begun to institute feedback systems by initiating performance evaluations of their buildings. For example, in the early 1990s, Public Works Canada launched a series of Asset Management studies followed up by "Serviceability Analysis" in which data on the technical and operational performance of buildings are collected through extensive interviewing of the building management team and building occupants. The National Buildings Agency of The Netherlands and the National Board of Public Buildings in Sweden both have systems for storing and accessing information about individual buildings, including the results of occupant surveys. The Swedish government also stores information on physical elements in public buildings based on input from senior and experienced technicians and from users.[16] Typically, in North America, businesspeople feel more comfortable investing in a computer tool that will accurately measure occupancy-related data in a building, rather than simply asking the people who work there. For example, the new indoor air quality monitoring system in Ameritech's 1.3 million sq. ft. building in Illinois collects data from 2000 VAV boxes and uses sensors to monitor gases and chemicals in the indoor air. Every six months air samples from throughout the building are sent out to a laboratory for analysis.[17]

Many data collection and analysis initiatives have been implemented in an effort to inform better decision-making in the design and construction as well as the operation of facilities. One international manufacturing corporation has developed a "Property Data Warehouse" for information about buildings and space, employee data, organizational data to connect data by divisions and cost center, and financial information by cost center, site and buildings. This information is made available to the company's business units to help them plan future space needs, monitor current space costs, and reduce space if needed.[18]

Several organizations have undertaken to analyze feedback from buildings occupants using a case study approach known as Post-Occupancy Evaluation (POE). This is an intensive analysis of information about individual buildings that emphasizes the functionality of spaces for workers over the technical performance of building systems. A large literature exists on POE, which has been implemented by federal agencies in the United States, such as General Services Administration and the Post Office, by large quasi-governmental organizations like the World Bank and IBM, and by numerous corporations in North America and Europe. Preiser lists published POE's from 1967 through 1986 on all types of buildings, including schools, universities, military installations, hospitals, housing, and offices.[19] But little is known about the strategies these organizations adopt to integrate the results into their decision-making process and apply the findings to improving business strategy. And gathering data from employees about their work environment is likely to be ineffective unless this activity is linked to ultimate applications of the results. As with all measurement, the challenge is not only to analyze and understand data, but also to be able to apply the newly acquired information to a useful purpose such as facilities problem solving and accommodation planning. To make occupants' feedback into an effective decision-making tool, managers must develop mechanisms for relating the results of data analysis to the decision-making needs of the organization.

TOTAL QUALITY AND CUSTOMER SATISFACTION

There are several well-established models available for making good use of users' feedback, and incorporating it into the ongoing decision-making processes of an organization. Each organization uses information in the context of its own culture, resources, and values, and feedback from users of buildings is nothing more nor less than information. Corporate personnel responsible for the O–A relationship need information to make good financial decisions as well as to be responsive to occupants' needs. Having managable, useful, and understandable information from building occupants about their environment also provides an opportunity for managers to open up communication with employees in an egalitarian and constructive way. By initiating an occupancy feedback system, with its attendant opportunities for positive and constructive user-manager communication, managers are in a position to exchange information about building performance; and just as feedback to employees about orders, deadlines, customers, and parts availability speeds up the manufacturing process, so knowledge about their building encourages workers to take "ownership" of their space, to be responsible for it, and to learn to use it to their advantage.[20]

For business strategists, the occupants of a building are the most valuable source of information about building performance, as long as the right kind of feedback is acquired. Merely encouraging employees to express their wishes and demands may sometimes be useful, but does not constitute measurement of the effectiveness of the accommodation. And measuring users' assessment of how responsive the environment is to their requirements is rendered more difficult and complex by the fact that the most appropriate work environment is, by definition, in a constant state of adjustment and alteration by occupants if they are really using it as a tool for work.

Using information from customers, especially feedback, to make good business decisions is one goal of the total quality management (TQM) programs that have been sweeping through North American businesses.[21] The principles of TQM were developed by W.E.Deming and are designed to encourage employees to think for themselves, to work in teams, and to strive for constant improvement.[22] Some of the critical steps in the total quality process include studying services and products for their weaknesses or flaws, aiming at continuous improvement, assigning control to individual workers, analyzing costs, instituting quality circles, "benchmarking" product quality as well as costs and time-cycles, and studying and improving customer satisfaction.

As facilities departments struggle to provide increasing levels of service for an ever shrinking budget, TQM programs have also become part of a new way of doing FM business. A key element of total quality for FM staff is measuring the quality of facilities services through feedback from their customers—namely the occupants of their buildings. But this is not systemic feedback as defined by Senge. It is "gathering opinions about an act"; as specified in TQM, measuring customer satisfaction is a valuable first step towards improving services. Along with attitude surveys, job satisfaction surveys, and a wide range of other employee-oriented "empowerment" exercises, many companies now send out customer satisfaction surveys to enable them to assess the quality of building services and space management.

Satisfying customers with good quality facility services is not the same, however, as acquiring feedback from occupants that can be applied to strategic accommodation decisions. The rationale for companies that initiate "client satisfaction" surveys is that building occupants are the clients or customers of facilities and building managers. The sign of a good facilities management team, therefore, like that of a good hotel or airline, is good quality service. In the role of service-providers, facilities managers are neither part of the corporate team nor engaged in business strategy; they are still providing a support role in the organization. Moreover, the customer satisfaction approach is weak from a strategic point of view because the results do not indicate how best to allocate resources to improve services. For example, if 65 per cent of respondents report themselves as "satisfied" with speed of response to complaints, or with indoor air quality, or with mail delivery, does this mean that services are adequate or inadequate? If they are considered adequate because over 50 per cent are satisfied, does this mean that some, a lot, or no resources should be allocated to improving the satisfaction of the other 35 per cent? How dissatisfied are the other 35 per cent, and how important is this information? What aspects of mail delivery or indoor air quality are they dissatisfied with? How is one to distinguish between "true" dissatisfaction and hard-to-please complainers?

In a total quality management program, client satisfaction surveys are designed to provide feedback to service-providers, as well as to improve communication between users and managers. However, using customer satisfaction with services as the yardstick for an accommodation strategy carries the risk of ignoring the difficult trade-offs that need to be made in space-related decision-making, such as number of people deployed in carrying out tasks, investment in equipment and new technology, current business goals, competitors' behavior, and likely value creation.

Measuring client satisfaction, while a useful beginning to more proactive facilities management, is therefore an incomplete strategy for managing the human element in modern office buildings. It encourages employees to think of their office space as they might think of a hotel room while on a business trip: "The company is paying so I want all possible services and conveniences while I am occupying this space." Office buildings cannot be run this way; it is not the building manager's job to provide unlimited conveniences and services to occupants. More suitable than the hotel room as a metaphor for strategically advantageous workspace is the serviceable computer. Computer users know what they want their computer to do, and in return for proper care and maintenance they expect it to do it. It is their tool, and its job is to enable them to work more effectively. They do not have expectations that go beyond the computer's capacity; if they do, they trade in their computer for another model. More importantly, in order to get the most out of the computer, they learn something about how the computer works, what it does best and what it is not good at, and they adapt themselves accordingly. Workspace needs to be seen as just such a serviceable tool, and using feedback from building occupants is one way that people can be encouraged to use their building to enable them to work more effectively.

In the table below, the distinction between satisfaction evaluation feedback and environmental assessment feedback is summarized.

Table 3.1. Distinctions Between Satisfaction Evaluation and Environmental Assessment

Satisfaction Evaluation	Environmental Assessment
Respondents list their wishes and unmet needs.	Respondents provide their judgments and perceptions.
Client survey implies promise to solve problems.	Occupants' points of view are acknowledged.
Results focus on improving customer satisfaction.	Results focus on improving work performance.
Occupants are passive recipients of facilities services.	Occupants and managers actively negotiate a quality environment.

This table illustrates how taking a total quality approach to measuring customer satisfaction is not necessarily effective in terms of the concept of "functional comfort": key to the distinction between customer satisfaction feedback and environmental assessment feedback. This is not to say that customer satisfaction surveys are ineffectual, but rather to point out limits on the usefulness of the customer satisfaction approach where accomodation strategy is concerned. What may work well in a TQM program for improving facilities services does not provide enough useful information to managers developing business strategy. Other approaches to feedback from building users, however, can, if carefully designed, constitute a strategic planning tool.

FUNCTIONAL COMFORT AS A STRATEGIC PLANNING CONCEPT

Functional comfort implies more than simple comfort of people in a building; it denotes the degree to which environmental conditions support building occupants in the performance of their tasks. The term "comfort" in some people's minds raises images of armchairs, soft lighting, and the impulse to kick off one's shoes and relax. Thus, comfort alone is not a concept uniquely conducive to the performance of work. The term "functional comfort" has been coined to define comfort in work-related terms, connected to the requirements of people's tasks. A functionally comfortable work environment *functions* to get work done by users as efficiently and effectively as possible.

Feedback on functional comfort is by definition oriented to decision-making applications. However carefully planned the work environment may be, changes in work-group configuration, or in the nature of the tasks themselves, can quickly render a place less than optimal. Unlike the concept of occupant satisfaction, therefore, which has no limits, functional comfort signifies a built-in indicator of when the desired state of environmental support for tasks at work has been reached. Whereas people can always become *more* satisfied than they are, they rate themselves as being functionally comfortable when the work environment optimizes the performance of their tasks. This is a fluid and changeable state, but it *is* a realizable end-state for a given time period.

Just as customer satisfaction can be measured to determine how well building services are being delivered to users, so functional comfort can be measured to determine whether the work environment is performing optimally for users. Feedback from building users is the most effective means of measuring functional comfort, whether it is in the form of POE's and similar initiatives, or whether simpler and less data heavy techniques are used. Functional comfort levels are typically assessed through occupant surveys, but, to be effective, the data must be related to the physical attributes of the work environment. Thus if 35 per cent of occupants are uncomfortable with lighting, it is important to have information on where these individuals are located in the building and what kind of work they are doing, in addition to their perceptual judgments. In this way, occupants' feedback on their experience of environmental conditions can provide a diagnostic analysis of building performance. The diagnostic approach has a greater impact on improving the efficiency of buildings as well as of employees.

One might ask why it is effective to ask employees for feedback on their work environment when their perceptions are so often colored by factors unrelated to the building. People have a tendency to judge their workspace not simply in terms of how it performs relative to their work, but in terms of offices they worked in previously (especially if they have just moved), the degree to which they like their job, rumors they have heard about the air quality or impending reductions in office size, and even whether it is spring outside, or summer or winter. It is, therefore, critical to manage the collection of feedback from occupants, and to use techniques of collecting information that focus on functional comfort issues. Furthermore, individual differences in job attitudes, seasonal changes, and corporate rumors have less of an influence on the data collected when users are addressed as a group rather than examining their individual comments and complaints. Many factors influence people's perceptions of their work environment, and it is ultimately those very perceptions that matter to managers. Managers are not doing research, nor are they trying to determine truth or falsehood—they are trying to improve the performance of work. Whether they are business or building managers, they are in the business of managing people, and it is ultimately the way that people see their world that will influence how people behave.

An essential element of the functional comfort approach is a mechanism to process, analyze, and use the feedback collected. Decision-makers may choose to act immediately on occupants' feedback, especially in those situations where employees' performance is obviously hindered by one or more environmental conditions. Or they may process feedback data more slowly, drawing on it to make far-reaching decisions such as whether or not to remain in a building. Even when the rationale for collecting feedback on functional comfort is clear, the application of results can still be a complex task, and how and to whom the results are disseminated in the organization has important consequences.

Functional comfort can be used strategically in an organization to evaluate the O–A relationship and develop plans for the future, to solve problems of employee productivity and morale, such as those incurred by sick building syndrome, and to resolve emotional conflicts around space-use issues. For example, in a research institution where teams of researchers work in laboratories, conflicts arise around space needed for expansion as research grants come in. These same teams resist all efforts to cut back on their space when research funds end or are cut. An effective way of dealing with this kind of territorial behavior is to introduce functional comfort criteria rather than simple square footage allocations into the debate. Instead of clamoring for "more space," occupants might provide a diagnosis of their environmental problems, which might include intrusive noise levels, lack of file storage, and insufficient privacy for concentrated work. This kind of feedback shifts the focus away from quantity to quality of space and helps resolve cost-related issues such as number of laboratories or critical laboratory size by encouraging users to focus on their workspace as a tool for work. In the case of the Bank of Boston described in the previous chapter, people were not directly questioned about their functional comfort, but manager-designer teams studied people's tasks and through an innovative process of their own devised functionally comfortable workspace that was strategically advantageous to the organization.

Both client satisfaction indicators and functional comfort ratings can be benchmarked to provide a context for evaluating the meaning of survey results.[23] In situations where feedback is used for diagnostic purposes, benchmarking can function as part of the diagnostic process. So when people in one building report discomfort, their assessment can be compared with people doing similar work in another workspace—not to ask "Am I better than you?" but to make the diagnosis of causes of

poor functional comfort more precise by referring to another, comparable workspace.

MANAGING FEEDBACK
FROM BUILDING OCCUPANTS

Functional comfort feedback can be collected in many ways. However,
regardless of measurement techniques used, the process of acquiring
and processing occupant feedback generally comprises five key stages.
These are:

1. collect information from occupants;
2. analyze and interpret the meaning of the information;
3. act on the results to solve problems and make decisions;
4. communicate with users regarding their feedback and follow-up actions;
5. negotiate environmental outcome with users.

A brief description of each stage follows.

1. Collecting Information

Information is collected by employing established social science methods and
market research techniques. User surveys, focus groups, key informant interviews, building walk-throughs, planning or building committees, and sugges-
tion boxes, are all techniques for gathering data. These techniques yield data
from occupants about their use of the space that can be analyzed: the data are
turned into useful information that finds its way to the individuals and groups
in the facilities organization that can and will make use of it in proactive decision-making.

   Once the data are collected and analyzed, occupant feedback can be integrated with information on the building's technical systems' performance, on
budget availability, on work-groups' expansion needs, on product changes, on
work-force shifts, and on other useful combinations.

2. Interpretation

The interpretation of data turns it into information. The data received from
building users are examined and turned into information about building
conditions: statistical results only acquire their meaning when they are related
to physical reality. For example, in a situation where 75 per cent of respondents
are critical of air quality, this could be due to odors, to over-heated space, to
dusty, dry or stale air. Additional measurements by instruments that count par-

ticulates, evaluate contaminants, and measure CO_2 levels may be useful at this stage.

If data on office location are collected together with occupants' environmental ratings, then the results might, for example, indicate more discomfort with air quality on floor 8 than on other floors in the building. This becomes useful information when conditions on floor 8 are examined only to find that an unvented print machine is operating on that floor, or that large numbers of paper files are changing hands there in the course of the day, or that there is simply a larger amount of desktop computer equipment on that floor. The focus of occupancy feedback is on understanding not just the *comfort* of occupants, and not just the building's technical *performance*, but on the *interaction* between users and buildings that is peculiar to each location of a work-group performing its set of tasks in a specific physical environment.

3.   Follow-Up Action

Acting on the results of an occupancy feedback initiative may mean assigning resources to solve an identified problem, but it may also mean collecting more information about the problem. A decision-maker may decide not to spend money on a particular problem at a particular time, either deferring it, or determining the building inappropriate for that degree of investment. Most commonly, however, acting on the results means using them to diagnose functional comfort problems in order to make decisions regarding problem solutions.

Different members of facilities teams need different information, and follow up in different ways. Feedback on lighting, noise problems and space use is useful to those responsible for design, construction or engineering projects. Real estate staff, responsible for site selection and lease negotiations, value information about temperature and ventilation conditions as indicators of mechanical system performance, and about lighting comfort, that can be applied to their criteria for the selection of space. Tenants use occupant feedback to establish conditions for leasing space and to negotiate solutions to problems with their landlords. Building managers want to know what priorities to set on building improvements and repairs that they are considering in order to be most responsive to users' priorities.

It is important to note that the implications of following up on functional comfort feedback from users may involve many levels of an organization in addition to facilities staff. This is one of the key ways in which facilities issues become integrated with the business operations of the organization. For example, indications of spatial problems resulting from insufficient file storage space may better be responded to through rules and practices about paper processing, including report production, document storage, and paper supplies, than by trying to increase amounts of file storage space. Such policies may be the responsibility of the head of a department, or of a vice-president of administrative affairs, or of the CFO. Paper storage often translates into big dollars when furniture standards, office size and ultimately rentable square feet are at issue.

In these ways, occupants' feedback is the business of the whole organization and is not limited to those responsible for buildings.

Follow-up does not necessarily mean a commitment to resource expenditure. If feedback is being acquired to test the success of a new furniture installation, for example, then a commitment to follow-up means understanding the results and incorporating the feedback into plans for future installations. In fact, follow-up action covers just about any action except ignoring the feedback once it has been requested. And yet, without adequate preparation, in large organizations, ignoring the results is often what is most likely to happen.

4. Communication

It is difficult to separate follow-up action from communication because communication must be ongoing as actions are taken in response to users' feedback, if the action is to be effective. Communication between managers and occupants starts at the data collection stage, when occupants are first informed about the purpose of their participation. The opportunity to provide data is itself a form of communication: in occupants' minds, managers are asking their opinions and giving them an opportunity to express themselves.

This does not mean managers have to act immediately to solve all problems that users have identified; some of the information can be acted on at once, whereas other results require lengthy planning before follow-up action is agreed on. When user feedback identifies problems that cannot be solved, managers also need to communicate the decision *not* to act, so that occupants have realistic expectations of what can and cannot be changed.

Figure 3.2 indicates the cyclical nature of the building occupant feedback loop. Managers' decisions affect employees; the employees, in their capacity as users of the building, provide feedback to managers about the impact on them of building decisions; decision-makers use the feedback information to make better decisions. Communication is valuable between building users and managers because of the opportunity it affords to negotiate a functionally comfortable work environment.

5. Negotiation

Negotiation refers to the process that management pursues if it wants its decisions about follow-up to occupants' feedback input to be effective. To initiate negotiation, decisions are communicated to occupants in a rational way with explanations of budgetary or other limitations on solutions to problems. Although this procedure sounds obvious in terms of how decisions might be taken in other parts of an organization, it is not usually applied to workspace issues. In fact, it is as unusual for facilities managers to communicate clearly with occupants about their building as it is for occupants to explicitly identify their job requirements to buildings staff. But if an occupancy feedback system is initiated, the effects of building-related information exchange and negotiation are:

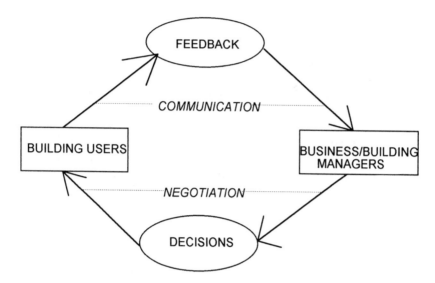

Figure 3.2. Cyclical nature of the building feedback loop, showing the relationship between users' feedback and managers' decisions.

- to encourage people to take responsibility for their own physical environment,
- to engage facilities personnel in the business activities and objectives of the organization, and vice versa,
- to teach coworkers to reach consensus regarding the environmental quality they need as a group, based on the *functional comfort requirements of their tasks.*

An effective occupancy feedback system sets in motion the negotiation of environmental quality between occupants and managers at the outset. The negotiation process is the most important element of the occupancy feedback system and it is a direct result of the quality of occupant-manager communication. Environmental negotiation with employees is a key mechanism by which facilities and real estate staff can engage in the business mission of the organization; and it comes about more effectively through eliciting occupant feedback on functional comfort than from measuring their level of satisfaction with services. Getting users to rate how well the office lighting enables them to do their work, for example, provides focused information that can be applied to negotiating with designers and managers the changes that need to be made to lighting, within the context of affordable technology and the limitations of the building (such as ceiling height).

Environmental negotiation does not work if managers use their information to repudiate employees' opinions. For example, complaints about thermal comfort may lead the facilities staff in a building to carry out instrument measurements in the problem areas ostensibly to determine technical solutions to the problem, but more probably to find out if the users are right. If thermometers show normal temperature and humidity readings, managers may be tempted to communicate this to the uncomfortable occupants as a way of informing them that the problem has been solved, or, worse, does not exist. But occupants' thermal discomfort does not disappear when they are told they do not have it. In fact, it may worsen: there has been no negotiation. To solve lingering complaint problems like this one, it is more effective to provide occupants with information about how their thermostats work, and the zones they control, to ensure that their desks and chairs are not placed up against large windows, and that no-one is seated under a ceiling diffuser that is blowing cold air. In some buildings, such problems are solved by enclosing thermostats or locking them, or even by disconnecting them and concealing the connected temperature sensors so that people may adjust the thermostats as much as they like without unbalancing the system and adversely affecting environmental conditions.

In summary, communicating and negotiating environmental outcomes with users of a building ensure that the decisions made on the basis of feedback information are effective. Environmental intervention based on the concept of the user and the environment as an interactive and dynamic system is more likely to solve environmental problems, to increase the work effectiveness of employees, and to reduce the costs of operating and maintaining buildings. In Senge's teams, this approach recognize the "reciprocal flow of influence" between users and buildings; it goes further than "gathering opinions about an act".

As the Bank of Boston example described in Chapter 1 demonstrates, an attractive and well-managed building attracts customers as well as well-qualified employees. To employees it can signal concern for their welfare, to clients a quality product, and to competitors it can signal success and a competitive edge. Well-managed real estate in a desirable location is a valuable asset and one that deserves to have resources invested in it if it is to make money for the organization. Problems such as indoor air pollution in an office building impede worker productivity by increasing time off the job, absenteeism, lateness, and, ultimately, staff turnover. Removing or at least alleviating such problems and increasing the quality of the building's environment increases organizational profit by reducing these costs. Much as companies like Body Shop International have substituted public support for environmental causes for conventional advertising, so companies like Bell Canada create a favorable image with their customers, investors, and employees by socially responsible property management practices: recycling waste, conserving energy, investing in the health and comfort of workers, and not consuming nonrenewable resources. Forward-looking companies are increasingly inclined to invest in their buildings in ways that enable facilities departments to operate as asset managers–"designing and maintaining a facility for competitive marketability"—rather than as cost centers.[24]

Failing to acknowledge to importance of the employee-environment relationship is not only inefficient and a wasted opportunity, it can also be expensive. Employees in one telecommunications company whose fears about indoor air pollution were not taken seriously, cost their company an estimated twelve million dollars when they evacuated two floors of an office building for seven days. Extensive testing and investigation failed to identify the cause of the odors that people reported, but the local health and safety enforcement body refused to allow the workers back inside the building. People went back to work after a week, but the cause of the odors was never found. This situation could have led to ongoing mistrust and an adversarial relationship between occupants and managers. However, on the occasion of the second walkout, more than a year later, building managers changed the situation by acting differently. They responded immediately with air quality testers, instruments, and advice; they moved occupants out of the offending areas; and they listened to each and every individual report about the supposed air quality problem. This time, there was no walkout; people were calmed and went back to work within 24 hours. The health and safety enforcers did not need to get involved, and the situation was saved. Building staff themselves attributed this avoidance of catastrophe to the rapid opening up of a clear line of communication between managers and occupants and to being available to discuss the problem with users before panic occurred.

Negotiating outcomes is more realistic that trying to create a perfect fit between employees and their workspace. As people's tasks change their knowledge and awareness and their work-group relationships, so do their environmental priorities; and most companies find it costly to keep replanning space and changing layouts. Replacing the ideal image of a perfect fit between users and their environment with a process of information exchange and negotiation that targets optimal functional comfort is a more cost-effective approach to space decision-making.

There is no single successful measure or final physical solution; there is only more or less money to spend, more or less manpower available, and a physical building which is not perfect. Successful communication and negotiation between building occupants and managers has a beneficial effect on the organization's bottom line; failed or nonexistent communication can be catastrophic.

Soliciting occupant feedback is the basis for environmental negotiation; responding to it is the next step; and managing the process of communicating results and negotiating follow-up actions to occupants is the third and crucial step. These "small steps" for the people involved in managing buildings can turn out to be a "giant step" for corporations in turning their buildings into worthwhile and long-term assets. In the following chapter, we examine one approach for eliciting feedback from occupants of office buildings that can be applied directly to turning the workspace into a tool for work. The Building-In-Use Assessment system measures occupants' functional comfort in terms of those elements of the physical environment that have impact an on occupants' performance at work and generates results that feed directly into the stages of follow-up, communication, and negotiation.

Notes and References

1. Veronique Robert, "Changez d'air," *Chatelaine* September 1993, p. 49.
2. Timothy J. Springer, "Does Ergonomics Make Good Business Sense?" *Facilities Design and Management*, July 1992, pp. 46–49.
3. MIT Laboratory of Architecture and Planning, unpublished case study, Cambridge, Mass. 1991.
4. John Markoff, "Apple Computer's new approach: private offices and common areas," *The New York Times*, 25 April 1993.
5. Buffalo Organization for Social and Technological Innovation (BOSTI) "*The Impact of the office environment on productivity and the quality of working life*" Buffalo: Westinghouse Furniture Systems, 1982.
6. J. Francis, D. Dressel, S. MacArthur, R.D. Neathammer, "Office Productivity: Contributions of the Physical Setting" U.S. Army Corps of Engineers, C.E.R.L. Technical Report P-86/13, September 1986.
7. Carole Sullivan, "Employee Comfort, Satisfaction and Productivity: Recent Efforts at Aetna," in *Promoting Health and Productivity in the Computerized Office: Models of Successful Interventions* eds. S. Sauter, M. J. Dainoff, and M. J. Smith; New York: Taylor and Francis, 1990. pp. 28–48
8. T. Springer, *Improving Productivity In the Workplace: reports from the field* St. Charles, Illinois: Springer Associates, 1986.
9. Architectural Research Centers Consortium, "The Impact of the Work Environment on Productivity" workshop proceedings, Washington, D. C., ARCC, 1986.
10. G. Raw, A. Leaman, and M. C. Roys, "Further Findings From the Office Environment Survey" London, England Building Research Establishment, 1989.
11. Kreon Cyros, personal communication, November 1990.
12. Springer, "Improving Productivity in the Workplace," p. 50.
13. "Improving the quality of light ... and work performance," *Office Administration and*

Automation May 1984, pp. 38–48.
14. Peter Senge, *The Fifth Discipline: the Art and Practice of the Learning Organization* New York: Doubleday, 1990.
15. Senge, *The Fifth Discipline*, p. 78.
16. "The Workplace Network: A Forum For Learning and Sharing" report on an international workshop Sweden, 1991, Ottawa: Public Works Canada, 1992.
17. "Indoor Air Quality: Blowing In the Wind," *Buildings* March 1993, pp. 46 ff.
18. Michael L. Joroff, "Corporate Real Estate 2000: Management Strategies for the Next Decade," The Industrial Development Research Foundation, 1992.
19. W.F.E. Preiser, H.Z. Rabinowitz, and E.T. White, *Post-Occupancy Evaluation* New York: Van Nostrand Reinhold, 1988.
20. See, for example, John Case, "The Knowledge Factory," *Inc.* October 1991, pp. 54–59; and E. M. Goldratt, and J. Cox, *The Goal: A Process of Ongoing Improvement* North River Press Inc., 1986.
21. "Quality: Small and Midsize Companies Seize the Challenge–Not a Moment Too Soon," *Business Week* November 30, 1992, p. 66
22. W.E. Deming, *Out of the Crisis* New York: Cambridge University Press, 1986.
23. International Facilities Management Association (IFMA) provides results from nationwide applications of its own client satisfaction survey, for example.
24. Christine H. Neldon, "Asset management: Benefiting facility management and the bottom line," *Haworth Office Journal* #2, March, 1992.

CHAPTER 4

BUILDING-IN-USE ASSESSMENT: AN OCCUPANCY FEEDBACK SYSTEM

"Failure to wring every benefit out of the most expensive capital asset most companies ever have would not be countenanced in any other aspect of corporate life"

John Seiler

HOW BUILDING-IN-USE ASSESSMENT WORKS[1]

Building-In-Use Assessment is a measuring tool to evaluate workspace from the point of view of the people using it. BIU Assessment produces data in a form that can easily be used by decision-makers to make changes to the workspace that they are sure will improve functional comfort and therefore the ability of workers to get their work done. Building-In-Use Assessment is not a *research* approach to the evaluation of buildings, like Post-Occupancy Evaluation. It is an *action* approach: it provides a measurement of occupants' perception of the physical environment in a readily understood and actionable form.

A company leasing some 30,000 square feet in a downtown office building was in the process of renegotiating its lease agreement with the property owner. The CEO recognized that some reduction in his staff might mean the firm was occupying too much space, but the landlord was offering very attractive terms for the same amount of space on another floor, including an offer to build out the space to the company's specifications. The CEO could not estimate the potential value to his workers of having a redesigned workspace—the current one had not seemed to pose any problems—nor did he have a tangible way of trading off the inconvenience and cost of a move with the potential advantage of better office space, especially in view of the uncertainty regarding future space needs.

To make the right decision for his company, the CEO decided to find out directly from his staff how well the existing space suited their work behavior. And rather than asking around informally if they liked their space and were comfortable in the building, he used a technique called Building-In-Use Assessment to enable him to measure which groups had enough space and which could use more, for file storage for example; which were working on computer screens and needed to be shielded from window glare and bright overhead lights; which were often out of their office and were amenable to shared workspace such as "hotelling"; and, finally, how those groups that might be losing staff might replan their space: larger individual workstations or more shared and team workspace.

The BIU Assessment process, in which data were gathered and analyzed, interpretations were provided, and follow-up action negotiated, gave the CEO the answers he needed. The process took a little longer than conventional space occupancy decisions do, but the cost was one-time and fractional compared to the differential over the long term between the two options the tenant was considering. On the basis of the BIU results, the CEO directed his administrator to negotiate a new lease on the same space, with an option to shrink it down within two years and with the landlord bearing the costs of space improvements. In carrying out these renovations to the space, the CEO was able to accommodate his work-force in a smaller more efficient configuration with the full support and approval of his staff who saw the effect of their input into the process, and who learned themselves through the BIU Assessment how to make better use of their physical environment as a tool for work.

The BIU occupancy feedback system employs organized and measured occupants' perceptions of their work environment as information to guide accommodation decision-making. Being systematic, rather than oriented to immediate problems and crises, the results of a Building-In-Use Assessment can be used to plan and allocate resources to building maintenance and repair, to replan and design office space, and to solve occupants' problems in cost-effective and manageable ways. Most important, a systematic approach to measuring occupants' functional comfort such as Building-In-Use Assessment encourages constructive communication between building occupants and managers which guarantees the usefulness of the information collected from occupants about their buildings.

More typically, feedback from occupants is in the form of individual complaints, and it is facilities managers or space planners who make decisions about space, not the CEO. The former usually do not have a systematic way of responding to occupants' complaints: aware of the shortcomings of their buildings, most are forced to struggle to find the resources necessary to maintain quality service to building users. Management-by-responding-to-complaints does not help managers budget for long-term improvements to offices, does not allow priorities to be set and met according to an established workplan, and does not allow business managers, like the CEO described above, to incorporate functional comfort into their business strategy. BIU Assessment is a system of managing feedback so as to inform management decisions, rather than reacting to individual complaints with piecemeal services and solutions. In this way, accommodation issues are linked with corporate strategy, and are ultimately related to strengthening the competitive advantage of business units.

AN INTRODUCTION
TO BUILDING-IN-USE ASSESSMENT

The BIU Assessment system is based on the following key premises.

1. It is only necessary to collect as much information as can be processed into follow-up actions and decisions, and no more.

2. The critical dimensions of the work environment to be assessed are those that affect the performance of work, which include those that affect team interaction and creative problem-solving.

3. A building's performance can and should be evaluated by its occupants in terms of the activities for which they use the building.

4. Feedback from users is not worth collecting if it is not integrated into an ongoing process of planning and action which guides change, solves problems, and aims at an overall increase in quality of the O–A relationship.

Designed originally to evaluate the quality of government offices in Canada, BIU Assessment uses a standardized occupant survey to collect data on seven key functional dimensions of the workplace environment, listed as follows.[2]

Air Quality: computed from ratings of air freshness, air circulation, and ventilation comfort, occupants also assess dryness of the air, whether there are odors, and if temperatures are too warm.

Thermal Comfort: computed from ratings of temperature fluctuation, and "temperature comfort"; occupants also assess if temperatures are too cold, if there are drafts, and, if temperatures are too warm.

Spatial Comfort: computed from ratings of furniture comfort, amount of space, and adequacy of work and personal storage space; occupants also assess access to meeting rooms and spatial layout.

Privacy: computed from ratings of visual privacy, voice privacy, and telephone privacy, occupants also assess amount of visual contact with the rest of the workspace.

Office Noise Control: computed from ratings of noise distractions, general office (background) noise levels, and specific noises of voices and equipment.

Building Noise Control: computed from ratings of noise from the air handling systems, noise from electric lights, and noise from outside the building; occupants also assess noise problems from office cleaners.

Lighting Comfort: computed from ratings of how bright lights are, whether there is glare from lights, and "electric lighting comfort"; occupants also assess if colors are pleasant, whether there is glare from windows, and if they have enough light.

The assessment works by comparing numeric ratings by occupants on each of these dimensions with normative scores in one of four Building-In-Use databases which comprise a total of some 7,000 respondents from over 30 office buildings.[3] The current database comprises scores from eight office buildings in the US and Canada, containing some 3,900 cases. Using a normative approach to benchmarking, the BIU database norms represent an average of occupants' ratings received from nine normal and unexceptional office buildings. They are not minimum scores from poor quality buildings, nor do they represent especially good quality space. The BIU norms provide a baseline of response to occupants' ratings of functional comfort conditions in modern office buildings against which new BIU scores from individual buildings can be assessed.

Feedback from occupants on these seven dimensions constitutes an assessment of their level of functional comfort. By relating these scores to physical environmental elements, such as floor layout differences, type of workstation, proximity to windows, air handling zone and other key locational attributes, a diagnostic analysis of the performance of the work environment for users is obtained. Unlike a conventional POE, BIU Assessment does not systematically investigate research questions, such as the impact of age, gender, job rank, and length of time in the building, on occupants' levels of satisfaction and comfort. And unlike the total quality feedback from customer satisfaction surveys, Building-In-Use Assessment does target the seven specific dimensions of work-space comfort to provide diagnostic, and, eventually, actionable information. Using functional comfort as a yardstick means that if people are uncomfortable in their work environment, it makes little difference to decision-makers whether it is due to their age, sex, or length of time in the building, but a big difference to know precisely the type and extent of the discomfort in order to be able to respond.

BIU Assessment is action-oriented rather than research-oriented in that its results target follow-up action and not reasons for people's behavior. Its analytic framework determines patterns of comfort and discomfort throughout a building based on users' experiences and perceptions. BIU Assessment in fact uses the building occupant as a sensing device: unlike a typical measuring instrument, the occupant does not need calibration. She works in the same environment every day of her life and is acutely aware of the ways in which the workspace environment affects her performance. Like Poe's "purloined letter", this obvious information is waiting to be gathered and used to improve her work environment, to improve her functional comfort, to improve productivity, and to make the physical environment into her tool for work.

Each of the BIU ratings incorporates the many influences on people in buildings, not by isolating and measuring each physical and psychological cause separately, but by making the individual user's own perception or judgment of the environment the data point. For some technicians, this focus on the user's experience feels like too much of a concession to subjectivity. However, the control exercised in an occupancy feedback system is the control of consensus: each individual's score is grouped for analysis, thus canceling out the effects of subjective differences. Moreover, as pointed out in the previous chapter, the effects of corporate culture, group values, and individual differences legitimately affect an individual's assessments because these conditions affect the judgments and perceptions of all employees and need, therefore, to be considered when interpreting the resulte and generating problem-solving recommendations.

The BIU survey can be initiated by anyone with a concern for health, safety and comfort in the office environment, whether a union representative, a CEO, a business unit manager, a facilities manager, or employees. The cost of distributing a short questionnaire, entering the data into a computer, and calculating the scores on the seven BIU dimensions is relatively insignificant. If it is connected with a renovation or a retrofit to part of an existing building, the cost may be 10 percent–15 percent of the price of a small renovation, dropping to 1 percent–5 percent of a large-scale building retrofit or new design. If the survey is simply an assessment of existing condition, and or associated with a design project, the cost can be controlled by defining how many users to survey and how much time to spend interpreting and acting on the results. The real costs of getting feedback from users is the introduction of a major new source of information into the existing decision-making system. More significant costs are likely to be incurred as staff take time to examine results, discuss what they mean, and decide what to do about them, especially if they decide to survey building users on a regular basis.

If Building-In-Use Assessment is used to develop a normative database for an organization, so that scores from individual floors or buildings can be compared to the norms for the organization, costs may be associated with database development. Some time and money are required to design and set up the database, to manage it so that it remains current, and to devise simple access software for designers, managers, and other decision-makers concerned about accommodation issues (see the first case study on BIU and database development in a large international organization, reported in Chapter 9). However, these costs should be evaluated against the potential benefits to the organization of having such a tool.

CARRYING OUT BUILDING-IN-USE ASSESSMENT

BIU Assessment uses a short survey questionnaire that is handed out to all building occupants. It is possible to sample a building and have accurate statistical results, but most companies prefer to involve all their employees in the survey in the interests of fairness and employee participation. The questionnaire comprises 22 rating scales of ambient environmental conditions with two additional scales to measure global judgments of comfort and satisfaction. Respondents are invited to write in "Additional Comments" in an open-ended format. Each item on the questionnaire asks the respondent to assign a rating between 1 and 5 in response to each question, where 1 is uncomfortable or bad and 5 is comfortable or good. A computerized calculation transforms the 22 ratings into scores on seven critical *Building-In-Use dimensions*. The scores on these seven Building-In-Use dimensions, also between 1 and 5, form the *Building-In-Use Profile*. While individual scores indicate occupants' assessments of individual work-spaces, the profile indicates occupants' assessments of functional comfort for the whole building. A summary of the technical underpinnings of Bulding-In-Use Assessment (including the BIU questionnaire) is provided in the Appendix.

The seven scores of the Building-In-Use Profile for a particular building or space in a building—once they have been received—are compared to the normative scores of the BIU database to benchmark them relative to normal office buildings in North America. The BIU database is newly constructed every few years, as data are collected from newer buildings. Company-specific BIU databases are also constructed from data gathered in that company's buildings so that organizations can compare specific buildings to their own stock. Table 4.1 shows how the normative scores in the BIU database have evolved.

Table 4.1. Evolution of Normative Scores in BIU Database

BIU Dimension on a scale of 1 to 5	First BIU norm 1984–88	Second BIU norms 1989–93
Air Quality	2.3	2.5
Thermal Comfort	2.8	3.1
Spatial Comfort	3.3	3.3
Privacy	2.3	2.9
Office Noise Control	2.9	3.0
Building Noise .Control	4.1	3.9
Lighting Comfort	3.3	3.6

There are two BIU databases from which BIU norms have been calculated and two proprietary BIU databases. The first BIU database comprises 2,800 respondents working in five Canadian government office buildings surveyed in the mid-eighties.[4] The second and current database, as indicated in Table 4.1 mentioned above, comprises scores from eight U.S. and Canadian office buildings surveyed between 1989 and 1993. These scores represent average-to-good quality private sector leased and owner-occupied office space in various cities and suburbs. Buildings surveyed in Europe and those with special and/or uncommon characteristics are not included in the calculation of norms in order to ensure that the BIU norms represent a typical or average North American office building profile.

The primary differences between the two sets of buildings from which these scores are drawn explain the shifts in the normative scores. In the case of the buildings in the first BIU database, computer technology was just beginning to be introduced. Many workstations had computer terminals, but not all, and hardly any had more than one. Printers were more likely to be impact (noise-generating) than laser (heat-generating). Some 30 percent of workstations were in perimeter locations and 70 percent in the interior of floors. There were more open-plan than enclosed offices in the sample. By the time data were collected for the second database, computer technology was fully introduced in most buildings, with the attendant problems of heat-generation, lighting and ergonomics. There are more corporate than government offices in the second database. Also, there is a higher proportion of enclosed offices relative to open plan in this sample, and, generally, more task lighting and window access to improve the lighting comfort norm.

The differences between the scores of the Building-In-Use Profile for a particular building and the normative scores from the BIU database comprise the *Building-In-Use Index*. An example of the results of a BIU Assessment (using the first set of BIU norms)is shown in Figure 4.1. Building X's scores are compared to the BIU norms; the BIU Index diag-

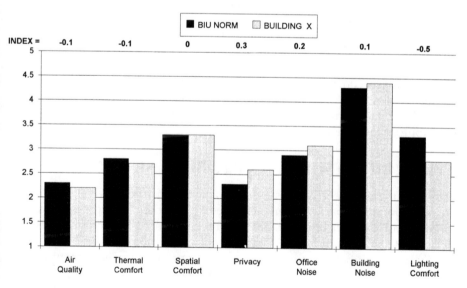

Figure 4.1. Example of a Bulding-In-Use Profile, showing the scores received from a building and compared to the BIU norms.

noses the relative positive and negative assessments of functional comfort in this building. In this example, lighting comfort, with an Index of −0.5, is the highest priority for needing improvement. Air quality and thermal comfort, with indices of −0.1, are almost normal. The best rating is on office noise control plus (0.2), which occupants find functionally comfortable: managers might do well to understand more about why this particular dimension is so successful in this building.

A building's BIU Index is the critical indicator to set priorities for follow-up action. On the basis of statistical analysis indicating the relative significance of each score's deviation from the norm, managers can target those items which require urgent intervention, and those which are questionable, adequate, and excellent. Statistical analysis of the BIU database demonstrates that, although all seven BIU dimensions are significant in predicting employee productivity, morale, and sense of health or well-being, not all are *equally* important in predicting these three behavioral outcomes. Table 4.2 shows how important each BIU dimension is in contributing to employees productivity, morale, and well-being.[5]

Table 4.2. Relative Importance of BIU Dimensions

Productivity	Morale (Satisfection)	Health (Well-Being)
Spatial Comfort	Spatial Comfort	Air Quality
Office Noise Control	Privacy	Thermal Comfort
Air Quality	Lighting Comfort	Lighting Comfort

Building noise control is the only Building-In-Use dimension not included in this table. This absence does not mean it is not significant, but rather that it is not one of the three most significant environmental contributors to these behavioral categories.

APPLYING BIU RESULTS TO PROBLEM SOLVING

Table 4.3, the stages of BIU Assessment are related to the five stages of an occupancy feedback system described in Chapter 3. The best way to condemn a report or a database of information to the scrapheap is to generate it without a clear idea of why it is needed. To be successful, a BIU initiative must include at the outset a clear statement of the way the information is to be used. On the other hand, in large organizations with complex decision-making processes, it is seldom obvious without careful study *who* will benefit most from occupant

Table 4.3. Stages of BIU Assessment Relative to Stages of Occupant Feedback System

Stage	Building-in-use Assessment
1. Collecting information	• Identify reasons why occupant feedback is needed • Identify who will be user/consumer of results • Distribute BIU questionnaire to all occupants • Ensure questionnaire is completed and returned
2. Analysis and interpretation	• Analyze questionnaire data and compute scores on the seven BIU dimensions of workplace comfort • Compare BIU ratings with database norms and compute BIU Index scores • Discuss results with facilities staff and technicians, employees and managers, and other interested parties: establish meaning of results in terms of actual environmental conditions of work

3. Action and follow-up	• Decide how to follow-up: decide who will take responsibility for implementing changes, acquiring resources, etc.
	• Develop recommendations for action: these include
	  - disseminating information about corporate policies or use of equipment
	  - retaining technical specialists for air quality sampling, lighting redesign, or other specialized intervention
	  - changing FM practices or policies
	• Develop criteria for eventual environmental improvement, e.g., ergonomic criteria for new furniture

4. and 5. Communication and negotiation	Implement sequence of communications with employees: 1. to acknowledge participation in survey 2. to provide feedback on results of analysis 3. to indicate changes or plans for changes 4. to keep employees informed on actions as they are implemented 5. to provide opportunities for employees to be aware of and have a say in actions taken.

feedback nor *how* they will benefit from it. Some companies insist on including in the process employee representatives from the spaces to be surveyed in the process by establishing them on a committee for the Building-In-Use Assessment. This group can then be involved in data interpretation by providing explanations and discussion of the results of the survey. This stratesy helps ensure that the survey is well-received by employees and that they fill in the questionnaire. It also helps to facilitate the interpretation of results as well as to manage employee expectations regarding follow-up action. Other companies, however, shrink at the thought of including employees to that degree; still others feel they would be unable to convince building occupants to dedicate time to something like feedback on building conditions when they have so much more serious work to do. In these cases, the organizational culture and norms of behavior indicate other appropriate routes and strategies to ensure that employees are as informed and involved as possible.

Interpreting BIU data is critical to the success of BIU Assessment. As well as expressing the goals and objectives of the occupancy feedback process in the corporate context, the data interpretation process is also a function of corporate culture and values. It can be structured, for example, as an inclusive process in which FM staff, middle managers, technicians, and employees all take part in discussions of results. Or this stage can be left entirely to expert consultants who connect the data from different parts of the building with physical conditions that can be measured in those areas and who can then generate technical recommendations for problem-solving and improvement. The most mileage to be gained in terms of improving communication, facilitating negotiation, and raising the level of functional comfort for occupants at least cost is likely to result from the more participatory techniques. People who understand more about their accommodation and how it affects their work are likely to be more open to a range of possible changes; whereas, a list of recommendations from experts often looks and feels like something that will cost too much money to implement.

During data interpretation, creative ideas for problem-solving and functional comfort improvements are generated. In turn, these creative ideas can be negotiated between users and managers, depending on the nature of the results and the degree of user involvement. In a recent, straightforward example, BIU survey results indicated noise discomfort in one section of the second floor. Interpretation revealed this to be attributable to a heavy fire door banging loudly every time someone used the fire stair. The facilities manager did not react defensively but was able to state publicly to building occupants that his efforts to solve the problem had so far failed. From feeling accusatory towards him, the employees shifted to an appreciation of his efforts on their behalf, and several useful suggestions were offered to try to solve the problem. Although occupants' level of physical comfort did not immediately improve, the positive experience of discussing a shared solution to the problem increased their mental comfort. They felt considerably less negative towards both building environment problems and towards building management.

Before building changes can be made to solve a problem, instrument measurements are sometimes required in order to determine the true source of discomfort. For example, survey data indicating air quality discomfort may reflect a range of causes that need to be measured before the specific problem in that building can be identified. Instruments are available to study airborne pollutants as well as number of air changes per hour, overheating, or other malfunctions of the mechanical systems. Unfortunately, as pointed out in the previous chapter, such follow-up measurements are often used, not to clarify and define a functional comfort problem, but to test whether or not users' complaints are *right*. As will be discussed in later chapters, getting feedback from users in order to prove to them that they are *wrong* is profoundly contrary to the spirit of the feedback exercise and contradicts the assumptions that underlie the functional comfort concept. This use can backfire for the organization. Using instruments to help interpret functional comfort ratings, on the other hand, helps to indicate problem locations and likely causes, so that follow-up actions can be specified. Useful directions for data interpretation are discussed for each of the seven BIU dimensions in detail in the chapters that follow.

COMMUNICATION WITH OCCUPANTS

Critical to the success of any follow-up strategy is good communication. Completing a survey form encourages employees to believe that their opinions are being sought for a reason, and that action will follow. To avoid bad feeling, it is advisable to let occupants know that survey results have been processed and actions are being considered as soon as possible after the completion of the survey. In some cases, managers fear informing occupants about survey results in case this causes occupant expectations to increase. However, as almost any good therapist will affirm, the first step towards effective communication is for one side to acknowledge what the other side has said.

It is usually not necessary to tell occupants which specific actions are being considered in order to establish communication and ensure that people do not feel that their input has been ineffective; it is sufficient to follow-up with some kind of response. Communication can and should begin before negotiation over specific actions. People's expectations of a survey in which they have participated are, *first*, that they have been heard and their input received; *second*, that something will be done to solve the problems they identified. They do not expect the solutions to

arrive immediately, but they do expect to see some movement in this
direction. It is not therefore always necessary to have solutions identified, financed, and approved before communicating with occupants.
The need for evidence of follow-up action becomes pronounced after
occupants have had repeated opportunities to express themselves and
managers have had ample opportunity to take action.

BIU Assessment generates recommendations for follow-up actions
ranging from small-scale low-cost changes, such as moving the desks of
people who report cold drafts from sitting adjacent to windows, to expensive long-term solutions like replacing workstation furniture with a
better and more ergonomically responsible design. Actions are more
likely to be taken to improve those BIU dimensions which receive negative ratings as these indicate likely or actual problems. Normal or positive ratings do not warrant corrective action, although a more detailed
understanding of why something is particularly comfortable or successful in a space adds valuable knowledge to inform future decision-making. Some of the actions recommended as a result of BIU analysis are at
a policy level and only indirectly affect the physical environment. For
example, where a company's paper storage problems have resulted in
rooms full of files and bookcases, in windowed office space used for archives, and have inhibited airflow due to high storage cabinets and
cluttered floors, a company-wide policy regarding limits on amount of
paper file storage permitted per person, or an investigation of replacing
paper with laser disc storage, might be recommended. In the case of
the CEO described at the beginning of this chapter, the BIU follow-up
was designed to clarify the situation for decision-makers, and only used
secondarily to help the space design process after the decision had been
taken regarding the future of the company's O–A relationship.

Typically, the approach to following-up on occupant feedback needs
to be planned as soon as BIU Assessment starts. Small, inexpensive
changes can be carried out immediately; others need budgetary approvals; and still others have to wait for the next fiscal year. Some are outside the province of building management and have to be passed on to
business managers or policy-making groups. In large organizations, an
action plan based on occupancy feedback has the greatest chance of success if it is developed by a team of individuals from different parts of
the company. This team approach also ensures the integration of accommodation with business strategy. Each action that is recommended
on the basis of the interpretation of BIU results can be reviewed, discussed, and expanded by staff, and finally shared with senior management and committees of building occupants. As well as familiarizing
facilities and space planners with the strategic concerns of each business

unit, such a process makes it incumbent on business managers and strategic planners to develop corporate accommodation strategy. This approach uses the assessment of employees' functional comfort as an integral element of accommodation strategy.

After a BIU Assessment has been completed, and environmental corrections have been made for immediate functional comfort improvements, additional surveys can measure the degree of improvement experienced by occupants, can compare space quality across buildings, can help set priorities on where to allocate resources to gain the greatest improvement, and can eventually establish an in-house standard of environmental quality for the workspace. An informed accommodation strategy incorporates BIU-derived action recommendations into a long-term plan for asset management of the company's space, equipment, and other capital investments. This plan, in turn, guides capital budget expenditures according to the priorities of the survey results while at the same time respecting both the constraints on resources available to management and the built-in limitations of building technology and performance.

BUILDING-IN-USE ASSESSMENT IN PRACTICE

BIU Assessment was used by two very different companies who wanted to build functional comfort assessment into their accommodation planning. The two examples described here show how corporate planners can use the qualitative parameters of functional comfort and, through BIU Assessment, quantify them to a degree that makes them useful to O–A decision-making. Although qualitative assessments based, as they are, on users' perceptions, may not always be accepted as a scientific approach, structured ways of incorporating people's feelings, hopes, fears, biases, blind spots and overall attitudes into decision-making are valued in organizations. Innovative management approaches incorporate these qualitative elements into planning in order to prevent human nature from sabotaging otherwise rational decision-making.[6] So while some technical types find BIU Assessment too qualitative because the scores are not produced by a piece of equipment whose measurement accuracy is demonstrable, other management-theory types find BIU Assessment too quantitative because it does not go far enough towards incorporating human nature and social processes in its approach. However, in practice, the in-between nature of BIU Assessment makes it a useful first step towards involving users in a dialogue about their workspace, as the two cases described below will show.

In these examples of how BIU Assessment works, key decisions were being made about space needs, occupancy, and the future of the company, and each organization used BIU Assessment to advance its understanding of its O–A relationship and to make decisions about its future that were not simply based on square footage calculations. The first example is a manufacturing firm that was facing drastic reductions in the number of government contracts it received and was looking to increase its efficiency. In the second example, a large law firm carrying out extensive renovations feared that the design process being employed would not take people's needs into account unless a deliberate effort was made to do so.

The manufacturing company occupying a campus-style set of buildings in the suburbs of Boston was looking to regroup its staff in fewer buildings to avoid layoffs and at the same time reduce costs. The facilities managers wanted to:

- establish the viability of the renovation and redesign decisions they had made in certain buildings,
- determine whether expensive mechanical upgradings in some of the older buildings were needed, and
- assess systematically the impact on the staff in one building of working in a windowless office space.

Facilities staff wanted to be able to use the occupancy feedback data to demonstrate to senior management that facilities dollars were being well-spent, and to show where and how employees could be regrouped efficiently without sacrificing productivity.

Employees in three buildings were surveyed using BIU Assessment. The occupants—highly trained professional and technical personnel with engineering backgrounds—were neither expecting nor were expected to have heavy involvement in the space planning process, but they responded well to a survey of their functional comfort ratings, expressing themselves freely and abundantly in the written comments section of the survey questionnaire. In all, three surveys were carried out. In survey I, data were collected from occupants of Building A, a newly renovated building which had been converted from an older building, with new spatial lay outs, new systems furniture, improved lighting, and an upgraded air handling system. This evaluation provided feedback to facilities staff on the success and correctness of their planning and design decisions in terms of the impact these decisions had on employees' experience of the work environment. Surveys II and III both took place in Building B: a single-story converted warehouse. Survey II was distributed while the space was windowless, shortly be-

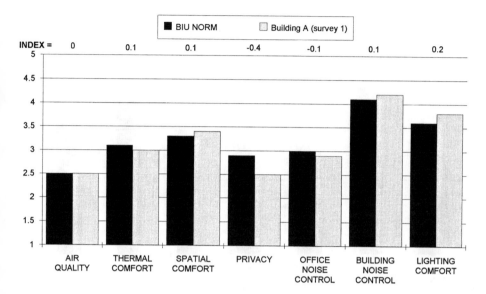

Figure 4.2. Building-In-Use Profile of Building A.

fore the renovations were implemented. Survey III took place some six months later, after windows had been installed and some other renovations completed. In this way, before and after results could be computed for the same group of occupants. The results of Survey I (shown in Figure 4.2) compared to the BIU norm show that the renovated work environment in Building A is close to normal on most dimensions, with the exception of Privacy. As a result, the facilities team felt comfortable recommending to senior management that the upgraded work environment of Building A be used as a standard for the quality of office space across the campus.

Figure 4.3 shows the comparison between the Survey II (before the renovation to Building B) and Survey III (after the renovation to Building B). These results show that adding windows, painting walls and changing the carpet improved not only lighting comfort and spatial comfort, but air quality and thermal comfort as well. Only privacy and noise control are lagging, but facilities staff were able to demonstrate to senior management that the $280,000 they had spent on upgrading the workplace had in fact measurably increased functional comfort; in other words, that people were working better in spite of the relatively modest amount spent.

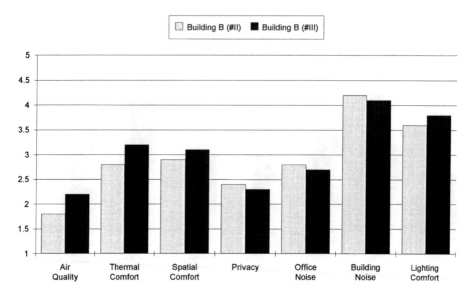

Figure 4.3. Comparison between BIU Assessment results from the first and second survey of Building B.

However, the figure shows that in spite of the increases in occupants ratings of air quality and thermal comfort, these are still below the BIU norms, demonstrating that although related renovations can have a psychological impact on perceptions of indoor air quality, occupants can still tell the difference between the appearance of improvement and the real thing. Therefore, in order to determine those areas in which the newly renovated Building B meets or fails to meet the company standard set by the renovated Building A, the results of Survey III (Building B after the renovation) were compared with the results of Survey I (Building A, the company's standard).

The results shown in Figure 4.4 indicate that Building B is less functionally comfortable than Building A on most of the BIU dimensions, except for thermal comfort and lighting comfort. Facilities staff used these results to demonstrate to management how much additional expenditure on Building B would be necessary to bring it up to the standard set by Building A. Moreover, if the money had been spent on a ventilation system upgrade in Building B, rather than on the installation of windows, it is not certain that the scores would have increased further than this. The facilities team now know they have gone as far as

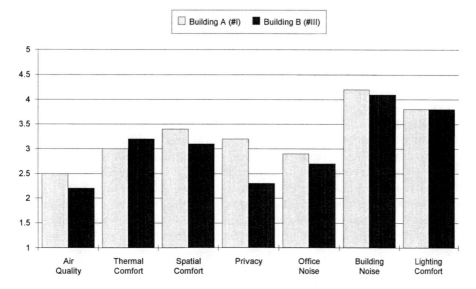

Figure 4.4. Comparison between BIU Assessment results from the renovated building and the company's standard (also a renovated building).

they can to improve occupants' comfort without physically upgrading the HVAC system. The company saved some $200,000 that they might have spent on a new air handling system—an amount which they may decide to invest in Building B if they continue to use it. Alternatively, with layoffs and downsizing reducing the size of the work-force, business managers may consider that there is no return likely from additional investment in that building and may adjust their accommodation strategy to regroup employees in those buildings which can perform more effectively as tools for work without upgrading.

The facilities team used BIU feedback results as the qualitative element in their closely-argued case made to senior management regarding future space use scenarios for the campus buildings. Instead of presenting simply dollar and square footage arguments for two equally good alternative options managers would have had to arbitrarily choose between, the team, by introducing easily measurable and comprehensible quality criteria, documented a strong rationale for an accommodation scenario that favored employees' ability to perform optimally in their space *as well as* an efficient dollar and square footage allocation.

In the second example, BIU Assessment was used differently, being a more traditional company with a more conventional problem to solve. Whereas the manufacturing company used occupants' feedback to make strategic planning decisions, the law firm saw it as a way to add value to conventional architectural services.

Located on 10 floors of a high-rise downtown office building constructed in the late 1970s, the law firm employs approximately 700 people. A decision had been made to renew the firm's lease with the provision that the 10 floors they occupied be redesigned and upgraded to the equivalent of a completely new office space. At the time of the Assessment, a nationally known architectural firm had been hired and work had begun on a major replanning and upgrade of the space. The facilities staff directing the effort determined that a structured opportunity for managed input from employees would help the architects, as well as have a beneficial effect on employee attitudes towards the considerable inconvenience, the moves, and the overall lowering of environmental quality occasioned by such a major project. Figure 4.5 shows the results of the Assessment of their pre-renovation office space, as compared to the BIU norms. Data interpretation of these results indicated, among other things, that air quality problems had been present since the first years of occupancy; that lighting was not effective for people working at computer screens; that offices shared between two

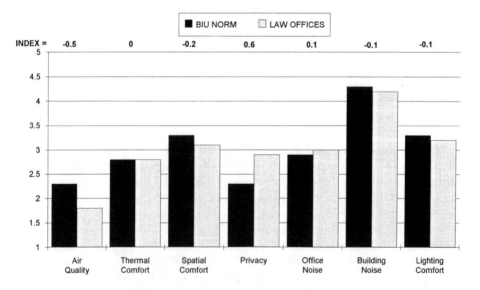

Figure 4.5. Building-In-Use profile of the law offices.

employees were uncomfortable; that some of the floors were crowded, generating several specific environmental problems; and that people in some areas were disturbed by the proximity of noisy office equipment. It also demonstrated that privacy was very satisfactory (important to attorneys) on most floors.

In order for this information to be integrated into design of the new space, the results of data interpretation were communicated to the architectural design team who were already at the stage of programming new spatial relationships and identifying users' needs. Of particular interest were the data from one floor which had been renovated the year before, using new furniture and a new floor layout. Both the client and the design team were interested in knowing if workers on this floor were more functionally comfortable than workers in the more traditionally laid-out floors because the renovated floor was planned in a more egalitarian fashion with only one secretary for two attorneys and more and better contact between secretarial staff members. It also had better lighting and more storage space for the secretaries. Figure 4.6 shows ratings from a typical floor (top) and ratings from the renovated floor (bottom).

These charts—where 0 means no difference between this floor and the rest of the building—show more positive ratings from the changed floor than from the more traditional floor. These results provided measurable support for the finding that the new prototype furniture layout was not only well-liked by users, but also rated better in functional comfort terms than the firm's standard layout on the other floors. The results were discussed and reviewed with the design team before being translated into a comprehensive series of design guidelines. The guidelines in turn provided the designers with information about ambient environmental conditions which would not otherwise have been available to them without extensive and time-consuming interviews and group sessions with occupants. The client directed the design team to adopt the new layout as their standard and to implement it throughout, thus saving time experimenting with still more new layouts and furniture systems.

From the clients' viewpoint, using systematic occupant feedback resulted in major savings of the designers' time. It enabled the design team to focus their programming inquiries on spatial adjacencies and the functional requirements of work-groups because the BIU information was available on environmental quality criteria. The client team later used the design guidelines as a checklist of users' documented functional comfort criteria for evaluating alternative schemes and proposals made by the designers. As a result, both the facilities staff who

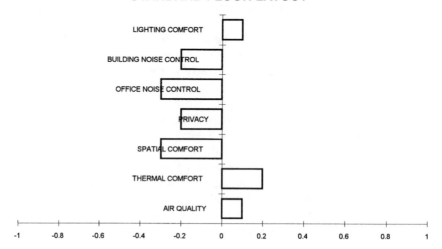

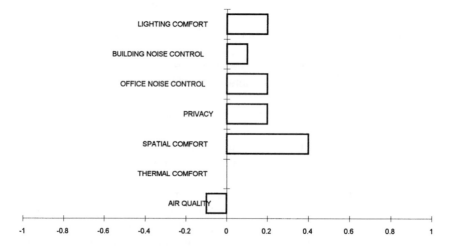

Figure 4.6. Comparison between two foor profiles in law offices, showing more positive responses on the prototype new floor.

represented the client and the members of the design team received enthusiastic cooperation from employees who felt that the BIU questionnaire had involved them in the process, that their opinions had been sought, and that their interests were safely represented in the decisions that were being taken. The law firm used the occupant feedback to negotiate a large-scale HVAC upgrading from their landlords, who had previously been unwilling to recognize the scale of the air quality problem experienced by occupants. The BIU results showing how inferior this downtown high-rise luxury office building was to national norms of comfort and well-being eventually forced the landlords to agree with the law firm that nothing less than a complete new HVAC installation was appropriate for the renovation.

On both occasions described in the examples above, occupant feedback was introduced into what might otherwise have been conventional space planning processes and made them unconventional. Was the outcome better as a result? The answer is that in defining qualitative bases for their space use decisions, both companies were able to avoid emotional territorial conflicts arising in the first case, from space reductions and group boundary redefinition, and in the second, from a move into less hierarchical and traditional space layouts. In both cases, employees were engaged in spatial decision-making through the BIU Assessment process.

USING FEEDBACK FOR CONTINUOUS IMPROVEMENT

To summarize, the actions taken by these companies in following up on their BIU Assessments were the following.

The manufacturing company has:

- vacated some of its buildings and accommodated a different config–uration of workgroups in the remaining space.
- selected which buildings to use and which to sell, lease out, or mothball.
- identified workgroups' needs in terms of their adjacency requirements as well as relative to building conditions and equipment.
- assessed the impact of moves on newly accommodated workgroups.
- programmed new space quickly and efficiently.

The law firm was able to:

- incorporate a complete overhaul of the office lighting into the renovation.

- implement an alternative office layout only in place on one floor on every floor in the new scheme.
- get agreement by the landlord to invest in a new HVAC system.
- negotiate a compromise with the designers to minimize the number of shared offices.
- help the designers negotiate with the client, on the basis of increasing functional comfort, the extra cost of installing glass panels in the interior walls of perimeter offices to permit more natural light to enter the space.

In both these examples, the significant expense of upgrading their accommodation provided decision-makers with an opportunity to invest in adding value to their environment by seeking out useful feedback from company employees in terms of environmental changes that are most conducive to the performance of work.

In table 4.4, summaries of the two companies' uses of BIU Assessment are presented in terms of the five stages of an occupancy feedback system that were described in Chapter 3.

The table shows that the impetus to collect feedback from occupants differed in each case. Although the survey approach used in both cases was the same, the analysis was carried out differently, with one company focusing on measuring the impact of their space-related decisions on occupants, and the other focusing on gleaning knowledge to guide the design of new space. In both cases, the feedback provided information that improved the space planning process and could be applied directly to improving each company's O–A relationship. In each case, the opportunity to provide feedback—to engage in communication and negotiation, even in a limited way—reduced the probability of territorial conflict, low morale, and user dissatisfaction, and helped defuse a potentially adversarial situation between building users and managers, thus ensuring that the decision-making process was not rendered ineffectual at a later (and more expensive) stage.

In both cases, facilities staff occupied traditional support roles. They did not seek to be considered part of the business planning of the firm. However, in initiating feedback from employees to aid in building-related decision-making, the facilities teams demonstrated the potential value of more strategically advantageous accommodation decisions. The manufacturing company's facilities managers demonstrated to corporate executives that although there was no obvious and short-term pay back (like avoiding a lawsuit) from improving the work environment, people's work performance could be improved by reducing the

Table 4.4. Five-Stage Summary of BIU Assessments by Manufacturing Company and Law Firm

Stage	Manufacturing Company	Law firm
1. Collecting Information	Facilities team initiated a series of BIU surveys of particular buildings or parts of buildings related to their expenditures on renovations and upgrading.	BIU data collected to help design team hired to replan new space as well as guide client's facilities team in selecting architectural schemes that corresponded to users' needs.
2. Analysis and interpretation	Analysis focused on assessing post-change impact or on pre- and postchange comparisons with an eventual view to determining most cost-effective interventions. Occupants not involved in interpretation of results.	Analysis focused on what to keep and what to drop from present environment for the new scheme. Client wanted to compare comfort ratings of occupants of prototype floor with ratings of other floors designed to firm's existing standard.
3. Action and follow-up	Actions included providing information to occupants, defining company's space standard, and using results to justify facilities expenditures to management.	Actions incorporated in a set of design guidelines that served as directive to design team as well as checklist for client to use evaluating alternative architectural schemes.
4. and 5. Communication and negotiation	Felt that the survey itself was sufficient and did not initiate additional communication with occupants. However, maintained the cycle of pre-change and post-change surveys so that occupants knew their opinions were being sought.	Used the survey to open up programming and space planning discussion between users and the design team. Feedback to employees became integrated with the cycle of the design process as interim architectural schemes were approved by staff.

adverse qualities of certain of their buildings. They demonstrated this impact by comparing functional comfort ratings from renovated buildings with those from unrenovated buildings. The law firm's facilities team saw the strategic advantage of involving employees in the design process of a major renovation, but wanted to avoid occupying their time in a lengthy user participation process. They also needed serviceable criteria to apply to their evaluation of the designer's proposals and recommendations and were able to use the BIU-derived design guidelines to establish critical environmental priorities according to users' functional comfort ratings. In both cases, facilities employees took it upon themselves to analyze the quality of the work environment in terms of occupants' requirements so that they could take steps to create an environment that functioned as a tool for work, although their mechanisms for doing so varied according to the corporate cultures to which they belonged. The net effect was the assumption by facilities staff of responsibility for the productivity of corporate employees in that they used occupant feedback to ensure that the work environment contributed to that productivity. This assumption of responsibility was an important first step in aligning facilities interests with the business goals of the organization.

In the following three chapters, each of the Building-In-Use dimensions will be discussed in detail. The discussions will provide examples of how BIU results have been interpreted and applied to problem-solving in specific situations. Three of the seven BIU dimensions pertain to building systems, and three pertain to the build-out of office interiors. Lighting Comfort is considered separately. The name assigned to each dimension has a specific meaning in the context of the Building-In-Use system, and each relates to certain typical challenges and dilemmas which arise in real buildings. In order for companies to define an effective accommodation strategy and derive full value from the O–A relationship, a creative response to these challenges and dilemmas is required, as well as a more profound understanding of how space affects people at work and how work-related behavior is affected by space.

NOTES AND REFERENCES

1. There are other published systems for evaluating functional comfort. See for example, *Performance of Buildings and Serviceability of Facilities* eds. G.Davis and F.Ventre, Philadelphia ASTM, 1990; and Rohles, F.H., Woods, J.E., and Morey, P.R. "Indoor Environment Acceptability : the Development of a Rating Scale" *ASHRAE Transactions*, 95, pt. 1, 1989, pp.23–27; see also Chapter 4 of F. Becker, *The Total Workplace*, (see Chapter 2).
2. For a complete account of the development of this system, see, Jacqueline Vischer, *Environmental Quality In Offices*, New York: Van Nostrand Reinhold, 1989.
3. A technical account of this process is included in the Appendix.
4. See *Environmental Quality In Offices* for details.
5. As reported in *Environmental Quality In Offices*, pp.146–152.
6. One well-known and widely-used example of such an approach is the "Managerial Grid", on which courses and seminars are available throughout North America, and about which several books have been authored by Robert R. Blake and Jane S. Mouton.

CHAPTER 5

BUILDING-IN-USE ASSESSMENT OF BUILDING SYSTEMS:

AIR QUALITY, THERMAL COMFORT, AND BUILDING NOISE CONTROL

"In many ways, air quality testing is like using a vegetable colander to trap a fruitfly that may or may not be floating in a bowl of milk."

Dr. Gemma Kerr

BUILDING SYSTEMS' DIMENSIONS OF FUNCTIONAL COMFORT

Each of the seven Building-In-Use dimensions presents an inherent challenge or paradox: none is what it seems once it is seen in the context of the user-environment system. These challenges are not insurmountable problems, but they are also not obviously resolvable using conventional analysis. Each requires what Handy calls "upside-down thinking" to understand it and know what, if anything, to do about it. In Handy's terms, upside-down thinking "has never been popular with upholders of continuity and of the status quo, ... invites one to consider the unlikely if not the absurd," considers seriously things "at first sight

91

impossible, or ludicrous," and believes in change and in moving forward into the unknown.[1] By understanding in a more practical context how these dilemmas and apparent paradoxes affect the user-environment system, all players in the system—managers, facilities staff, building users, designers, and builders—can learn to direct a closer and more informed eye at the environmental quality of different types of workspace.

The BIU dimensions of *air quality, thermal comfort* and *building noise control* are connected because they depend on the operation and functioning of building systems, primarily mechanical and air handling systems. Air quality presents a sort of generic mismatch—a credibility dilemma—in which information gathered from occupants about the nature of their air quality experiences in the workspace often fails to match information about air quality conditions based on data collected through conventional instrumentation. We call this the Air Quality Dilemma. Another BIU dimension, thermal comfort, presents a quirky paradox: BIU ratings of thermal comfort for a space, as derived from a Building-In-Use Assessment of that space, have been found to bear no obvious systematic relationship to the number and type of hotline complaints building managers have received about temperature problems for the same space. And building noise control poses a different sort of problem: using instruments to measure noise levels may tell us if background noise levels are too loud, but will not give us equally important information, namely that occupants are uncomfortable when the background noise levels are too soft.

Taken together, these three dilemmas provoke interesting and informative questions about functional comfort as it relates to building systems. Are surveys of users measuring inappropriately and should they be abandoned in favor of other more reliable indicators of the status of these three dimensions in a building? Or do people simply not always tell the truth? Or perhaps different measurement techniques fail to complement each other and have the effect of counteracting each other's results? A deeper exploration of the Air Quality Dilemma may assist us in answering these questions.

THE AIR QUALITY DILEMMA

On receiving indoor *air quality* complaints from building occupants, managers typically opt to carry out instrument testing, either to have a better definition of the problem, or to find a solution to it, or to find out whether the complainers are imagining things. The results of such tests often show that the air meets all existing health and comfort standards, thus placing managers in a significant dilemma. If complaints have been voiced, it appears responsive for building managers to hire air quality testers and determine whether or not current standards are being met in the indoor air, or, as it is often put, to verify whether or not "we have a problem." Usually in such cases, samples of air are drawn randomly and measured for their level of a number of contaminants (e.g., formaldehyde, CO and CO2, respirable particles, and Volatile Organic Compounds). It is not unusual for the testers to return to their client with a report that shows the levels of these contaminants to be well below acceptable limits, therefore concluding that the air is not polluted, the complainants are mistaken, and the managers do not have a problem. [2]

In fact it is at this point that the manager's problem gets serious: if occupants report headaches, fatigue, nausea, and skin rashes, and they fear an indoor air quality problem, their fears are unlikely to be allayed unless their symptoms go away. If their symptoms do not go away, the information that their air meets current industry standards is unlikely to reassure them. It may even make them suspicious. Their complaints usually continue, and managers still have a problem, often one that consumes staff time, requires a constant outlay of resources, and reduces people's effectiveness on the job.[3] The dilemma does not only occur in office buildings: complaints from flight attendants about symptoms resulting from a presumed reduction in amount of fresh air circulating through the cabin during flights resulted in measurements being taken which showed no violation of accepted indoor air quality standards. The matter was taken up by Congress and the Centers for Disease Control in view of the continuing anxiety of the flight attendants.[4]

Why do occupants' perceptions differ from instrument measurements? There are many reasons. One is that indoor air testing is far from an exact science: it has grown up over the last fifteen years, instrumentation is still developing, and standards of acceptability for pollutants are far from complete. Testing often focuses on air samples drawn from random locations throughout the building and can be limited to once or twice over a given time period. Each analysis of the air sample to identify possible pollutants and their relative levels in the sample adds to the cost of the testing and often yields incomplete information. Existing standards are inadequate in that threshold limits for contaminants, although well-authenticated, are only available for a fraction of existing chemicals, and those that do exist may not protect workers against chronic exposure.

Other reasons for the discrepancy between indoor air quality (IAQ) measurements and users' perceptions lie in the psychology of the workers themselves. People at work in a building are affected by numerous environmental conditions, of which the contents of the air is one. Others include temperature, chair comfort, and overhead lighting, and these conditions do not impact the occupant separately but interact in their effect on users. Added to these are the less tangible effects of ignorance about how HVAC systems work, anxiety about indoor air pollution based on what they've read and heard, the quality of the relationship with building managers, the effects of organizational culture and communication, and products and tools that they may be using at work. In the context of this large user-environment system, it is impossible to isolate a few instrument-based numeric measurements of some components of the indoor air to explain the wide-ranging and complex set of behaviors that constitute human discomfort at work. While not dismissing the value of taking measurements, it is reasonable also to address these complex behaviors, and to do so in a way that acknowledges all these influences on human perception. In fact, people will often report indoor air quality problems when they themselves cannot identify the source of their discomfort at work.

Occupants' Perceptions of Air Quality

Where does the Air Quality Dilemma originate? Everyone has ideas about indoor air quality: few people in North America are completely unaware of the health and comfort issues related to mechanical ventilation in sealed buildings. Office workers' fears and concerns are expressed in complaints about odors, stuffiness, dry air, and physical symptoms of ill health. For building managers, ensuring high-grade mechanical system performance, testing air regularly, and maintaining recommended standards of ventilation are among their highest priorities. And a veritable industry of chemists, engineers, laboratories, instrument manufacturers, and air quality experts has grown up in just a few years to respond to the perceived or real threat of polluted indoor air in modern office buildings, especially in structures considered energy-efficient, where no outside air leaks in and relatively little air from outside is circulated through the building.

Industry experts approach the measurement of indoor air quality in two fundamental ways. One is taking air samples and looking for the presence of contaminants in the form of chemicals, bacteria, or molds; and the other is analyzing the mechanics of HVAC system performance, and ensuring that fresh air, temperature, and other critical variables are performing to specifications. Most investigations of IAQ problems involve physical analyses of air handling system performance as well as testing for possible pollutants. Agencies who sponsor air quality research to determine acceptable standards of contaminants include National Institute for Occupational Safety and Health (NIOSH) and American Conference of Governmental Industrial Hygienists (ACGIH) whose standards for commonly measured pollutants such as formaldehyde, ozone, carbon monoxide and carbon dioxide are widely used when air samples are drawn for contaminant testing. The American Society of Heating Refrigeration and Air Conditioning Engineers (ASHRAE), on the other hand, has developed a widely used consensus standard for ventilation system performance, based on amount of outdoor air provided to each person in the building.[5]

From the occupant's viewpoint, poor indoor air quality is most often perceived in stale air, odors, stuffiness, warm air, dry air, and physical symptoms such as headaches, nausea, and fatigue in the afternoons. In some cases, employees insist that their symptoms do not improve until they leave the building. Occupants are also aware when groups of employees are apparently infected by such diseases as colds and flu which, they believe, have been transmitted throughout a building by its air handling system. Most people believe their air quality would improve if they could open windows; they are usually ignorant of how the ventilation system works, and many believe that in sealed buildings the same air is endlessly recirculated.

People's negative opinion of indoor air quality is reflected in the Building-In-Use norms. The norm in the original BIU database for air quality is 2.3 on the 5-point scale, and in the newer database it is 2.5, signifying that people's perceptions of indoor air quality (and therefore the air quality itself) have improved between buildings built in the early eighties and those built towards the end of that decade. In spite of this improvement, however, on a scale of 1 to 5, these are not high scores. The fact that 2.5 is normal does not mean it is good: it means that overall in office buildings across the continent, building occupants judge themselves as more uncomfortable than comfortable with this aspect of their work environment. However, in newer buildings, the BIU ratings for air quality are improving, especially where some effort has been made to provide really good indoor air quality. For example, a score of 3.5, or one entire scale point above the BIU norm, was the rating in accountants' offices located on three floors of a relatively new, high-rise luxury office building in Boston where significant retrofit had been carried out by the tenants—and by the owner on behalf of the tenants—mainly along the lines of installing free-standing air conditioning units. A similarly high score was also received from one of the newest buildings on the campus of Harvard University where a top quality air handling system had been specified to ensure the best possible indoor air quality.[6] It was gratifying to determine through these high ratings that occupants in both cases were clearly aware of and benefiting from the expensive efforts that had been made on their behalf.

Building-In-Use Assessment of Air Quality

Because occupants' ratings of the key dimensions of their environment are integrated into a single system of inquiry (the Building-In-Use Assessment system), occupants assess *air quality* in the context of their other functional comfort experiences. This distinguishes Building-In-Use Assessment from other occupant surveys which focus exclusively on a single ambient condition, for example, ventilation or temperature. By rating all seven environmental conditions simultaneously, occupants can indicate where their experience of *air quality* comfort, for example, fits with other possible sources of discomfort, such as lighting or noise. One of the effects of the elevated public awareness of likely threats to health from the air inside sealed buildings is that the first thing employees think about when they experience uncomfortable physical symptoms, such as headaches and sore throats, is the indoor air. Occupants' complaints about air quality alone can mask other sources of discomfort. In fact, the BIU Assessment carried out in Building G, the profile of which is shown in Figure 5.1, illustrates a possible explanation for the Air Quality dilemma: perhaps managers have trouble identifying measurable physical causes of indoor air pollution through testing the air because occupants' evaluation of indoor air quality is a catch-all category that flags any type of functional discomfort in the building.

In Building G—a large, 15-year old owner-occupied, multistory, suburban office building—environmental problems in previous years had created expectations among employees of a serious air quality problem. In spite of corrections, people were still complaining. The low ratings received on almost every one of the seven BIU dimensions do indeed show that much more than indoor air quality is bothering these workers. As well as a low air quality index of −0.5, both thermal comfort and lighting comfort have index scores of −0.4. In addition, spatial comfort, privacy, and noise control are all rated as dysfunctional by the occupants of this building. It is clear that the air quality problems of previous years have left their mark on these workers, creating an unresolved psychological problem in their attitudes towards the building. Data interpretation indicated that lighting comfort was probably the most serious source of discomfort in the building, and that oc-

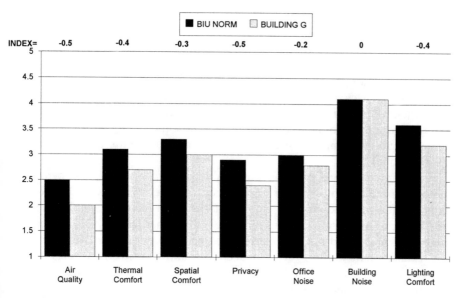

Figure 5.1. Building-In-Use Profite of Building G, showing discomfort on almost every dimension.

cupants' ratings of air quality and thermal comfort were likely reflecting symptoms resulting from eyestrain and visual discomfort rather than ventilation problems. These employees worked extensively on computer terminals, and the building's lighting was designed to accommodate office work in the 70s. Occupants' symptoms—headaches, nausea, and feeling too warm—were clearly attributable to the inappropriate brightness and glare conditions. An overall lighting retrofit to the building was not as expensive as another overhaul of the mechanical systems, and the data analysis indicated it would be twice as effective in reducing occupants' level of discomfort.

As with all the BIU dimensions, the BIU air quality score incorporates the many influences on people in buildings and therefore provides a more profound analysis of likely sources of discomfort. In terms of the Air Quality Dilemma, results of the BIU Assessment suggest that using instrument testing alone to define and/or solve air quality problems is something of a non sequitur. When people complain about indoor air quality, who knows if they are referring to indoor air, or if they mean indoor air plus a number of other environmental conditions, or if they are not referring to indoor air at all. In fact, BIU Assessment can ensure that instrument measurements are useful by calculating BIU ratings of air quality for different locations in the building, such as each floor, and different mechanical zones. Through these calculations, it is possible to define with more precision what kind of indoor air quality problem is actually occurring, and where. Even a building which receives a satisfactory air quality rating on a building-wide basis may show specific areas (a basement location, or near an elevator shaft) where air quality scores are low. In such cases, indoor air quality testers brought in by managers to solve the problem can efficiently focus on those areas represented by the low ratings, and, through understanding some of the elements in the particular micro-environment of those workers, can define the best type of measure to take.

For example, a group of uncomfortable occupants may work with fume-generating equipment or paper; they may be close to a leaky exhaust shaft from a kitchen or garage; or the fresh air dampers for the air handling unit serving their space may be jammed closed. However, these findings would not necessarily become apparent through conventional instrument testing; occupant feedback allows instrument tests to be very precise and targeted. This amounts to a staged approach that integrates instrument measurement with occupant assessments in a mutually beneficial way. The Air Quality Dilemma has a better chance of being solved by such an integrated approach than by the mutually exclusive "my data [air sample analysis] against your data [survey questionnaire]" approach which generated the dilemma in the first place.

Analyzing occupant feedback systematically informs managers more precisely of the seriousness and extent of discomfort than do the telephoned-in complaints and individual calls for service generated by perceived air quality problems. Occupant feedback shows whether there is a generalized indoor air quality problem, or whether a small group of overly sensitive individuals are reacting to unfounded fears of pollution in their work area, or whether some other environmental factor is causing the discomfort.

THE THERMAL COMFORT PARADOX

From the occupants' point of view, the most significant aspect of their *thermal comfort* in office buildings is its *variability* and its *unpredictability.* Some days are too warm, others are cold; or mornings are cold and afternoons too hot. This variability may occur in time, varying throughout the day or the week, or in space, varying as people move across a floor or between floors. Paradoxically, a standard of thermal comfort exists, based on the results of instrument measurements, that is widely used but does not predict or take into consideration temperature fluctuation in time and space.

The term "thermal comfort" is most commonly used to designate the ASHRAE thermal comfort standard which defines a thermal comfort envelope within which 80 percent of occupants are comfortable. The envelope comprises measurements of mean radiant temperature, ambient temperature, relative humidity, and air speed. It can also incorporate clothing and activity levels. It usually translates into indoor temperatures of 68° to 72° F in most office buildings, with relative humidity levels (Rh) of 30 to 40. Typically, with the heat generated by equipment, people and lights, managers find it more difficult to keep buildings cool than to keep them warm, and during harsh, extremely cold winters, relative humidity in many buildings typically drops down closer to 20.

People at work dislike experiencing temperature extremes in their workspace, and they dislike not knowing how temperatures will fluctuate and not being able to do anything about the fluctuation. If people are trying to concentrate on work, being too hot or too cold intrudes on task concentration. As a result, they are quick to complain to building managers. Studies show that although occupants do not have the same level of concern with thermal comfort as they do with air quality, individual complaints about temperature are the most frequent source of complaints in office buildings.[7]

The Paradox of Thermal Comfort is that occupants' functional comfort—in terms of Building-In-Use Assessment—cannot be explained *either* in terms of the ASHRAE standard, *or* in terms of the complaints that occupants make by telephone or in person to building management. In fact, case studies of thermal comfort problems show no connection between numbers of calls received from building users who are too hot or too cold, and the thermal comfort ratings they provide in their BIU Assessment. This is paradoxical, because it would seem that if degree of reported discomfort were linked systematically to number of individual complaints received in most buildings, the BIU thermal comfort norm would be down around 2.0, like air quality. The fact that it is not suggests that, in spite of individual complaints, when thermal comfort is assessed along with other environmental conditions in the context of the overall impact on getting work done, it is not as functionally uncomfortable as air quality; temperature is just something people find it easy to telephone in complaints about.

The Thermal Comfort Paradox may be partly explained in the following ways. When comparing service call volume with systematically collected feedback, there is an inconsistency in how the timeframe for the data is defined. It is difficult to know what period of time is being judged when occupants rate thermal comfort in their workspace, whereas numbers of service calls must be counted within a specified time period. It is also possible that occupants are quick to initiate complaints about thermal comfort because they believe temperature to be easily controlled, adjusted, and changed, so there is a good chance of getting action on this particular complaint. In this sense, thermal comfort problems are far more actionable than acoustic or lighting issues and generate a demand from occupants for corrective action in a way that other elements of the work environment do not, even though they may be causing more discomfort. Complaints that are initiated by occupants, or by one occupant, have to be reacted to by facilities staff whether they come from six people or six times from the same person. On the other hand, in initiating a request for feedback from occupants through a questionnaire or other type of survey, managers control the form in which they receive this information, can ensure that one rating corresponds to one person's experience, and can put the individual ratings together to get a picture of the whole. As with air quality, using a systematic approach to occupant feedback ensures that thermal comfort ratings are integrated with ratings of other aspects of comfort in the work environment, and can therefore be analyzed relative to other likely sources of discomfort.

Building-In-Use Assessment of Thermal Comfort

Thermal comfort complaints should not be dismissed as unimportant, yet before describing thermal comfort as the most significant problem in modern office buildings, we need to evaluate what causes people to complain about temperature. Figure 5.2 compares Building-In-Use Assessment ratings from the same offices in early spring and in the summer. The BIU norm for thermal comfort is 3.1. It used to be 2.8 in the database comprising 1980s office buildings, in which computers were just being introduced with no special provision for their heat generation. The results shown here are from a newly-built and newly-occupied office building. Feedback was sought on how functionally comfortable it was for occupants both immediately after moving in (March) and a few months later (June), in order to compare the two.

   The comparison shows that many of the ratings dropped slightly as people settled into their space and become more accustomed to it—overcoming the halo effect of the move to new space. However, the biggest single drop between March and June was in thermal comfort, with a difference of −0.4 between spring and summer. Remembering

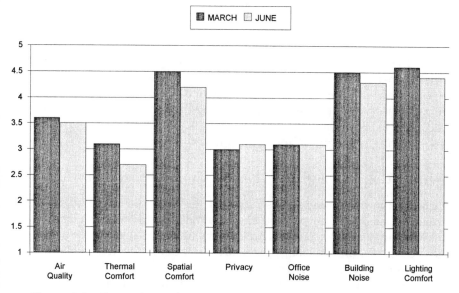

Figure 5.2. Comparison of BIU Assessment results from the same building surveyed in March and later in June of the same year.

that the space was only occupied in February, occupants had not experienced this space in summer when they rated it in March. With such a distinct pattern of discomfort, it is certain that building managers will be able to trace a problem with the air handling system's temperature control that emerges under warm climatic conditions. Data interpretation showed the discomfort condition to be more pronounced at locations near windows. An examination of thermostat location showed that the zones controlled by each thermostat were located at some distance from it, and the system therefore could not maintain the ASHRAE standard effectively. The managers used the BIU data to negotiate some retrofit work on thermostat location and functioning from the building owners.

As with air quality, instrument readings of thermal comfort taken in functionally uncomfortable areas of a building help identify excessively high or low temperatures and ensure that there is a correspondence between the sensor readings that control HVAC performance and actual conditions in the space. In another building, BIU Assessment identified a particularly low thermal comfort score on one of the three floors surveyed. When data-logging temperature and humidity recorders were placed throughout the floor, it was determined that in some areas temperatures climbed to over eighty degrees Fahrenheit two or three times a week in the afternoons, only coming down to normal levels after workers had left for the day. A check by the landlord revealed a malfunctioning ventilation unit in the ceiling which was rapidly repaired. On the other hand, some building managers respond to reports of thermal discomfort like Mr. Brown, facilities manager for a large publishing company. When occupants of his building complain of uncomfortably cold temperatures, he brings his thermometer into the space and takes a reading. "See this!" he shouts triumphantly to his long-suffering occupants, "You're warm!".

Thermal Comfort and the Thermal Comfort Standard

Mechanical systems are specified to meet the ASHRAE standard when a building is designed, but the specifications are based only on estimates of the density of people and equipment anticipated for the space. These estimates are speculative to start with, and, over the lifetime of the building, these densities change. Only comprehensive technical measurement can indicate whether the ASHRAE comfort standard is being maintained over time. Feedback from occupants suggest that in many buildings, it is not. In older buildings, constructed in the seventies and even early eighties, no provision was made for the heat generated by computer equipment because the proliferation of so much powerful office equipment was unanticipated. Even in new buildings, mechanical system specifications turn out to be inadequate, because tenants or owners are increasing equipment and reducing space standards to accommodate more people in less space.

As space uses change, some owner-occupiers examine their mechanical systems' specifications to determine whether or not they are still capable of handling the amount of people and equipment in the building. Users' complaints about fluctuating and unpredictable temperatures usually indicate that system alterations and adjustments have not kept up with space use changes. As a result, ASHRAE's thermal comfort standard may well prevail for part of a day or a week, but unpredictable fluctuation of temperature and humidity conditions means it is not being maintained consistently. Although widely used, the ASHRAE thermal comfort standard does not address the variability problems that affect users' functional comfort. And thus it is that building managers can maintain the ASHRAE standard and yet receive a constant stream of individual complaints about temperature conditions.

Thermal comfort problems are often traceable to the location, accessibility, and performance of thermostats. Expecting control, people in offices fiddle mercilessly with thermostats; they expect to control temperature as simply at work as they do in their homes. But thermostats are often inappropriately located with regards to the zone they are controlling; thermostats for one room may be located in another. Moreover, in adjusting these thermostats, occupants have little understanding of the technology in the ceiling or under the window that the thermostat controls. They do not understand climate zones and the effects neighboring zones have on each other. Managers throw up their hands in despair at the proliferation of thermal comfort complaints, lock up the thermostats, install dummy thermostats, or blame the one or two individuals who complain most frequently and vociferously. To the managers, the thermal comfort standard is being maintained, and people are making a fuss about nothing. One way to improve thermal comfort in office buildings, therefore, might be to upgrade thermostat technology, either by providing thermostats with which individuals can control their own environments with full instructions and a map of the zone they occupy, or to enclose, lock in, or otherwise remove thermostats from access by occupants so that they cannot even try to adjust their own environment.

There is no reason to infer from BIU Assessments either that the ASHRAE standard is ineffectual in maintaining users' thermal comfort, or that complaints and service calls from building occupants should be ignored. However, the Thermal Comfort Paradox means that the standard is hard to maintain consistently, and that therefore people will always complain about being too hot or too cold as long as they believe that something can be done about it. Unfortunately, people are reluctant to take responsibility for the impact of their own actions on thermal comfort. Playing with thermostats, wearing inappropriate clothing, and moving one's chair into an inappropriate location, such as next to the window, are all common ways in which people sabotage their own thermal comfort and proceed to blame the building. The best way to respond to the Thermal Comfort Paradox, therefore, is to educate users to take more responsibility for their accommodation. Facilities managers who just react to complaints are not helping users learn about their physical environment; whereas proactive communication and negotiation between users and managers will increase the awareness of building users and ultimately improve their functional comfort.

THE QUANDARY OF BUILDING NOISE CONTROL

The third of the three Building-In-Use dimensions that relate to building systems is *building noise control*. This dimension is distinguished from the office noise control dimension, which refers to noise generated

by coworkers or related to coworkers, such as keyboard noise and telephone conversations. (office noise, control will be discussed in detail in
the next chapter.) Sources of building noise are more remote and less
easily identifiable. They are related to the operation of the building:
the noise caused by ventilation and electrical systems, and, less frequently, noise from the lights or plumbing, and noise from outside the
building. People's tolerance threshold for building noise is higher than
for office noise, in part because the sources are more impersonal.

The Quandary of Building Noise Control faces space planners and
designers as well as those who would measure building quality. Managing noise is not a simple matter of reducing noise to as low a level as
possible. The quandary lies in the fact that too little noise is as functionally uncomfortable as too much noise, and that good noise control
means providing the right noise level for functional comfort, which is
not necessarily the least noise. The BIU noise dimensions cannot simply
reverse the comfort scale and claim to be measuring the "silent-ness" of
a workspace. Noise in the work environment, especially that which is
generated by machinery and remote sources, can *increase* acoustic comfort by masking people's voices and other intrusive sounds, or it can *decrease* comfort by drowning out conversation and generating stress.
Thus the scale of noise comfort in the work environment ranges from 5,
which is a comfortable level of noise control, to 1, which indicates poor
noise control and, therefore, an uncomfortable noise situation caused either by little or too much. The quandary lies in the difficulty of speci-
fying solutions to a noise problem that may be caused both by the
presence and by the *absence* of noise.

Prevailing acoustic standards do not acknowledge this bidirectional
problem. They focus on protecting hearing in industrial environments
from loud noise and on reducing stress from noise. They address being
able to hear conversations while at the same time protecting
confidentiality. They also, to some degree, concentrate on reducing
stress. Some instrument measurements of building noise in offices focus on measuring background sound levels (such as the rumble of mechanical systems when the building is unoccupied) to determine the
degree to which they are likely to disturb the acoustic conditions required for effective office work. Formal acoustic standards as such are
not well-developed for office environments. In many office buildings, a
background hiss or rumble from the mechanical systems is noticeable
but rarely intrusive. In fact, such sounds can be positive for occupants
when taken as a sign that the ventilation is functioning, especially in
buildings where they are suspicious of the indoor air quality, so ventilation systems which cycle on and off can also generate anxiety in build-

ing occupants by seeming to indicate reduced air flow. For occupants of open plan or cubicle-type offices, the background noise from air systems functions as a sound-masking system even though air handling noise is not appropriately calibrated to mask human voice frequencies. Air handling noise provides the illusion, at least, of masking conversations, and this causes people to feel they have increased privacy. Thus a silent air handling system would not only cause people to think there was no ventilation, but would also reduce their voice privacy and increase their level of discomfort in open plan offices.

Building-In-Use Assessment of Building Noise Control

The normative score on building noise control in the first BIU database is 4.1, and in the second it is 3.9. Both scores are relatively high on the 1 to 5 scale, indicating that typically this type of noise is not a problem in most of today's office buildings. This means that either the noise levels generated by building-related sources are not disturbingly high (if it is disturbingly low, this is reflected in the office noise control score because of the lack of background noise to mask office sounds), or it means that whether they are high or not, the occupants tolerate them well, better in fact than occupant-generated office noise, which has a lower BIU norm. It is easy to overlook the contribution of building noise control to the functional comfort of employees because it is not a widespread source of stress and discomfort, nor is it a critical element in managing the work environment. But research on environmental stress indicates that sustained noise and vibration generate physiological stress that can impact employee performance and morale over time. Thus when occupant feedback on building systems is evaluated for follow-up action, adjustments and overhauls to the mechanical systems need to take the noise aspect into consideration. Case studies indicate that in those rare cases when building noise is a problem, it is very detrimental to the performance of work.

The graphs in Figure 5.3 compare two floors in a building where the building noise control score was not noticeably poor in the overall building's BIU Profile. The floor differences show that localized noise sources can affect people's functional comfort in their immediate surroundings, even though the scores may cancel each other out when the whole building is examined. Differences between the two floors turned out to be due to the presence of retrofitted return air fans that were lodged in the ceiling of one floor, and were very noisy when they were turned on. However, when they were inactive, there was insufficient air movement on the floor. The higher, and closer to normal, score on

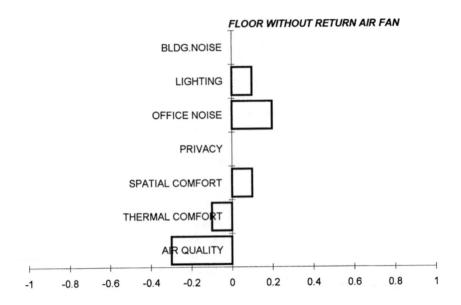

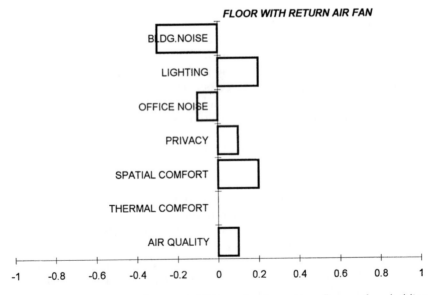

Figure 5.3. Comparison between BIU results from two floors of a bulding demonstrating different noise conditions.

building noise control, therefore, is connected to uncomfortable air quality and thermal comfort; whereas the low (uncomfortable) score on building noise appears with a better score on air quality.

There are many possible causes for intrusive noise from air handling systems which affects occupants adversely. What they have in common is that they are almost all difficult to correct or solve. For example, in office buildings with individual heat pump units above the ceiling, each unit emits a constant fan noise as well as a more intrusive noise as the compressor cycles on and off. In new buildings, where unit maintenance and repair has not fallen behind, this noise is mildly annoying and tends to rate only slightly below the norm. As such a system ages, noise problems increase and occupants' functional comfort decreases. In buildings with centralized air handling systems, noise may result from an imbalance in the air delivery system. Dampers may have been closed off above the ceiling (perhaps responding to thermal comfort complaints) causing more air to be forced out of diffusers in other zones. In other cases, like the one described above, where an additional fan unit has been installed above the ceiling, usually as a retrofit measure, occupants seated below are impacted directly by its noise. In this building, this situation caused workers to turn down their thermostats in an effort to shut off the noisy ceiling unit. This action in turn caused a reduction in their air flow and a decrease in air quality as well as decreased thermal comfort as temperatures rose.

Low (uncomfortable) building noise control ratings often come from buildings with individual fan coil units mounted on the walls beneath the windows. These units are installed because of the opportunity they can offer occupants for individual control over heat and cooling, or to increase the amount of cooling available to interior offices. However, when in operation the noise of the fan combines with other office equipment noises such as the drone of a disc drive or the hum of a laser printer. The background noise level in enclosed offices where this situation has occurred is such that workers cannot hear to engage in telephone conversations—a powerful cause for reduced functional comfort.

A not infrequent building noise problem is caused by rumbling and vibration of the fan systems themselves, often located on the roof. People working on the top floor of the building are disturbed in a way that is imperceptible to a casual visitor. In one building, vibration from the roof-top systems was significant enough to make bathroom doors rattle in their frames, thus generating additional noise in neighboring offices. Systems which have major air intake units and exhausts on the roof and provide air mixing and distribution individually on each floor also generate poor building noise control scores from people seated in the vicinity of these mechanical rooms, each of which contains a substantial amount of vibrating equipment.

Managing Building Noise Control

Building noise control is at the other end of the actionability scale from thermal comfort when it comes to the possibilities for effective intervention by managers. None of the examples offered above lends itself to rapid, inexpensive, or effective correction. There are no cheap and effective strategies to effect noise reduction on heat pump compressors, for example. And to reduce low-frequency noise and rumble from roof-mounted systems, the equipment (usually mounted on dense rubber pads) has to be structurally isolated from the physical structure of the building. One building that received a very low building noise control score was found to have not only noise from heat pump compressors above the ceiling, but also noise from outside the building. As luck would have it, the building was located beside the commuter rail tracks, and high-speed trains went past several times a day, vibrating windows and disturbing people at work. The only solution to this problem would have been moving to another building—an expensive solution indeed.

Where occupants are disturbed by low frequency noise and rumbling from fan systems either on the roof or in mechanical rooms on each floor, the resources are not usually available for the complete insulation of the equipment from the structure which would be necessary to alleviate the problem. Managers need to evaluate the significant expense of such an intervention against the increased ability of employees to perform well once their functional comfort has been improved. In one building where executives were located at the top of the building, building noise control ratings indicated uncomfortable noise levels both from the roof-based fan systems and from additional retrofitted air conditioning units that had been installed in mechanical rooms on the floor itself. But even for their executives, managers had not undertaken to spend the significant sums of money required to improve their building noise control, noting both that the executives were not often present in their offices and that it was "only their secretaries" that had to deal with the adverse environmental impact.

In environments where there is insufficient noise, correction is also expensive. Most obvious is the installation of a sound-masking system in which boxes above the ceiling emit randomized noise at appropriate frequencies to mask the human voice. It is not obvious that this is an effective solution in the long run, since the background hiss can also drive up noise levels, as people speak louder to be heard, therefore causing office noise control problems. As building noise control is *not* one of the top three functional comfort dimensions to have an impact workers' morale and their productivity(see Chapter 4), and it *is* one of the most costly types of problem to solve, cost-effective and value-creating follow-up to solve the building noise control quandary is hard to define. Building noise is most amenable to follow-up action when only a few people are affected by a problem—such as noisy diffusers in one part of the ceiling—and their functional comfort can be improved cost-effectively by adjusting ventilation equipment.

HOW BUILDING SYSTEMS CAN INCREASE FUNCTIONAL COMFORT

Occupancy feedback studies in modern office buildings suggest that HVAC technology in such buildings is often of comparatively poor design and/or quality. Badly-performing HVAC systems are in part the result of major changes in office space use over the past decade, with some ten times the amount of heat-generating equipment now in offices than was anticipated when most office buildings were built, placing ever-increasing demands on the building's ventilation systems. Together with energy management concerns, more high-rise buildings with sealed windows, and the widely reported prevalence of sick building syndrome, concerns of both tenants and owners have generated a demand for a more sophisticated HVAC technology that is slow to emerge. Most North American building stock was, of course, built more than ten years ago, and many of the problems encountered by today's workforce result from aging HVAC equipment with insufficient capacity trying to cope with the increased heat load of lights and equipment in the modern office. However, it is disappointing to examine the HVAC technology in newer buildings only to realize that, although ever increasing quantities of fresh air are being cycled through buildings, there is relatively little innovation in the way these building systems are engineered. For example, user control over temperature and ventilation is poor, humidification technology is often disappointing, contaminants are inadequately evacuated, and there are still dead zones of stale air, temperature imbalance, and noisy equipment to be found in areas of most modern office buildings.

Moreover, modern and sophisticated technology is only as effective as the skills and knowledge of the people operating the equipment. As

HVAC technology, especially climate controls, becomes more sophisticated, so building technicians take on more complex responsibilities. Managers who are not technically trained themselves assign control over the mechanical systems to one or two individuals on the facilities team who may or may not understand the technology they are controlling. Unchecked by their supervisors, who understand even less than they do, their mistakes go unnoticed and possible solutions to HVAC problems are ignored. This is compounded by the fact that the engineers who design the HVAC systems are not those who install them, and neither of these actually operate or necessarily service the systems. As a result, equipment operators become dependent for information and assistance on how to operate the systems on the organization that has the servicing contract. This situation effectively removes the locus of responsibility for good HVAC performance from the consultants who specified the system as well as from those that installed it.

In a more future-oriented scenario, building occupants would be instructed in HVAC system performance, so that they too can take responsibility for optimal functioning of building systems. Air quality problems such as odors and thermal comfort problems caused by thermostat malfunction, often arise as a direct result of occupant ignorance. Managers can overcome occupants' ignorance by making employees aware of the effects of such simple acts as taping over diffusers, storing files on top of window units, placing large pieces of heat and/or fume-generating equipment on an open office floor, storing food in their desks, and storing and handling large quantities of (particle-generating) paper. As HVAC systems become more sophisticated, it is increasingly important for building occupants to take some responsibility for their effectiveness. In buildings with unsophisticated or inadequate HVAC systems, it is important for managers to inform occupants when their expectations of HVAC performance are unrealistic in view of the technology installed in the building. Such actions ensure that information about the building is communicated to users as part of the organization–accommodation feedback loop described in the previous chapter. Users need to learn about their building just as managers need to learn about how people use the space.

In summary, improving workers' functional comfort in relation to building systems has come up against three unique challenges. In each case, although the way occupants assess these three areas of functional comfort can be measured and understood, occupants' perceptions fail to correspond *either* to traditional disciplinary definitions of areas like air quality and thermal comfort, *or* to be fully understandable and measurable in those terms. Commonly accepted measurements of indoor air quality often fail to find that standards have been violated, even when occupants are uncomfortable and report illness and discomfort; thermal comfort is the most common complaint in modern office buildings, but all evidence shows that it is not the most uncomfortable condition—except when the ASHRAE thermal comfort standard cannot or does not apply; and building noise problems can be caused by too little as well as by too much noise, with the result that little is known about their long-term effects on occupants: they are also likely to be costly to correct. In each case, knowing precisely what to do or how to respond to occupants' feedback is complex and requires innovative, "upside-down," thinking. It also requires an open line of communication between occupants and managers and some form of environmental negotiation. Strategic decisions that pertain to occupants' functional comfort in relation to the operation of building systems are capable of responding to these challenges by applying some of the solutions described in this chapter as well as by making more explicit demands of building systems technology.

As a greater understanding of user behavior in buildings develops, so the quality of buildings themselves, and particularly of building systems and HVAC technology, will become more responsive to human needs and to the real-life way in which people at work behave. In the next chapter, we move on from the three functional comfort dimensions pertaining to a building's mechanical systems to another three BIU dimensions, namely those that address building interiors and issues of furniture arrangement, spatial lay-out and noise control in the office.

NOTES AND REFERENCES

1. Charles Handy, *The Age of Unreason* (Boston: Harvard Business School Press, 1990). pp. 23-29
2. Jacqueline C. Vischer, "Using Occupancy Feedback to Monitor Indoor Air Quality" (Proceedings of ASHRAE IAQ Conference, June 1993); *ASHRAE Transactions*, 99 pt.2 (1993).
3. Jacqueline C. Vischer, "The Psychology of Architecture," *Los Angeles Times* 28 March 1988; sidebar: "Bateson Building: Engineering A Better Workplace," p. 4.
4. "All Things Considered," *National Public Radio* 30 July 1993.
5. ASHRAE Ventilation Standard 62–89.
6. Constant Volume Dual Duct system.
7. *"Office Tenants Moves and Changes,"* Building Owners and Managers Association, (Section 3) (Washington, D.C.: BOMA International, 1988).

CHAPTER 6

BUILDING-IN-USE ASSESSMENT OF PLANNING AND DESIGN OF INTERIOR SPACE: SPATIAL COMFORT, PRIVACY, AND OFFICE NOISE CONTROL

"Space and spaciousness are closely related terms, as are population density and crowding; but ample space is not always experienced as spaciousness, and high density does not necessarily mean crowding."

Yi-Fu Tuan *Space and Place.*

FACTORS INFLUENCING THE DESIGN OF SPACE FOR WORK

The three functional comfort dimensions of *spatial comfort, privacy,* and *office noise control* are connected because they pertain to the way space planning locates workers in interior space. Issues of enclosed versus open plan workstations, spacing and size of workspace, file storage and accessibility, space standards, furniture, and height of partitions are all relevant to occupants' ratings of these three dimensions. As space managers and designers know, there is nothing like office size, work storage and filing space, and partition height to generate controversy, conflict, and sometimes outright bad feeling among office workers because people associate these factors with their status and rank in the organization.

The greatest threat to functional comfort on these three dimensions is overcrowding. A crowded office environment generates poor ratings on privacy and office noise control as well as on spatial comfort. A feeling of crowding is not always clearly related to objective measures of density or even to numbers of people in a space: some job requirements and workgroup orientations encourage close working proximity in which employees tolerate densities that elsewhere would be experienced as crowded. For some types of work, a degree of crowding is considered desirable, for example, in laboratory research, where the open and abundant exchange of information contributes to the creativity and performance of researchers. Thus having to line up to use a piece of shared lab equipment, while annoying to users at the time, can actually have the effect of increasing opportunities for informal communication that would be missed if everyone stayed in their own lab. However, there are also non productive crowded situations, where noise, heat and smells from co-workers slow people down and lower morale.

As with the functional comfort dimensions related to building systems, discussed in the Chapter 5, there are unique dilemmas associated with the functional comfort of interior workspace planning. And as with those dilemmas, some "upside-down" thinking is required to resolve these issues as well. For example, spatial comfort—one of the most critical of the BIU dimensions in terms of occupants' productivity and their morale—has generated a sort of crisis in modern offices. Restrictive space and furniture standards are proliferating in organizations as space planners try to fit more people into less space, but a potential crisis has been reached in the amount by which individual workspaces can be shrunk down. The Spatial Comfort Crisis is related to the Privacy Conundrum, which states that although workers say they need individual privacy and complain if they are deprived of it, more and more office work is carried out by *groups* of people, rendering the concept of *individual* privacy somewhat obsolete. And third, the Challenge of Office Noise Control—which is linked to the privacy conundrum—is the design of appropriate enclosure to control office noise effectively; this is quite a challenge in the open office and in group workspace.

In this chapter, each of these dilemmas will be analyzed in terms of type of problem and likely solutions encountered through Building-In-Use Assessments of actual buildings. Creative ways of addressing spatial comfort, privacy, and office noise control are key components of successful functional comfort at work.

THE SPATIAL COMFORT CRISIS

In the absence of comfort standards for workspace layouts, like the ventilation and thermal comfort standards developed by ASHRAE, companies have hastened to develop their own corporate space standards to regulate individual office and workstation size, degree and height of enclosure of individual workspace, and type and amount of furniture allocated to each person. Large dollar expenditures on furniture systems are often rationalized to management by a hoped-for reduction in amount of space per employee, and therefore a reduction in overall amount of space occupied by the organization. But although *individual* workspace is shrinking, shared workspace and group areas are increasing. And, as larger amounts of space and equipment are needed for support functions such as meeting rooms, copy centers, libraries, lounges, lunchrooms, smoking rooms, and cafeterias, the overall amount of square feet per person is not necessarily declining. Owing to growing office equipment requirements, the amount of space per person has steadily increased over the past ten years, averaging out in North American office buildings at between 200 and 300 square feet per person.

Part of the Spatial Comfort Crisis is generated by employees themselves, who are often not aware of a shift in space allocation from individual to group workspace and are frustrated and angry about the perceived shrinking size of their office or workstation. And in spite of planners' efforts to develop space standards that correspond to employee functions and task requirements, in many companies space is still assigned on the basis of rank and status differences. Typically, managers and above are in the larger, enclosed offices, often with windows, and technical and support staff are in workstations or cubicles with less than full-height partitions. The more traditional the organization—like, for example, law offices—the more likely managers and professional staff are to have both full enclosure and window access while executives get these as well as the largest offices. This leaves support and technical staff located in the interior of floors, often in smaller quarters, in spite of the fact that they are there more of the time, they have more equipment, and more of it in use, and that the organization's revenues are often dependent on their level of productivity.

Reference was made in Chapter 2 to the virtual office which uses computer and communications technology to allow employees to work from anywhere outside of, and often in addition to, the company's office building. One of the driving forces behind the push towards virtual office use is the savings anticipated in real estate costs. In one corporation, managers estimate that on a typical day in their headquarters accommodating almost 5,000 people, 10 percent of the desks are unoccupied because people have not been hired yet, people are sick, on vacation, or working elsewhere. Another 10 percent may be vacant because people have left their desks for meetings or tasks in another part of the building. As a result, at any one time, 20 percent of the workspace might be vacant, making it imperative that managers start rethinking individual workspace allocations to make their accommodation more cost-effective. Some companies are doing away altogether with conventional office space, encouraging people to work how and where works best for them.[1]

BUILDING-IN-USE ASSESSMENT OF SPATIAL COMFORT

The Building-In-Use Assessment approach addresses more than simply the amount of physical space; occupants assess workspace layout and type of furniture, circulation and meeting areas, degree of enclosure and proximity/separation of workgroups. Managers' *spatial comfort* decisions have a major impact on dollars invested in such costly items as furniture, carpeting, and partitioning systems. Property managers and building owners are often reluctant to get involved with their tenants on these issues, even when the tenant has no inhouse expertise in workspace planning. Many companies rely on their design consultants for these decisions. For owner-occupiers, staff responsible for space planning and floor layouts are likely to be different from staff who specify furniture and staff responsible for purchasing, and these again may be different from those responsible for indoor air quality, HVAC performance, and occupants' thermal comfort. The result of such fragmentation is that major investments are made on the basis of incomplete information about space needs, furniture appropriateness, and the functional comfort of interior layouts.

Measuring spatial comfort includes measuring:

- ergonomic dimensions of people's work environments, such as chair height, back and seat design, and comfort.
- height and depth, and often number, of work surfaces.

- placement and adjustability of computer screens.
- amount of storage space employees have for files and work-related documents as well as for their personal effects.
- opportunities for both formal and informal meetings and for working in groups.

The norm for spatial comfort is 3.3 on the 5-point scale. The norm is the same in the first, government building BIU database as it is the second, private sector building BIU database, probably because the floor lay-outs and furniture do not vary significantly in the different buildings. The spatial comfort score is therefore predictable across a large range of types of building interiors, including open plan, partitioned office space, and enclosed offices. The stability of this score also suggests that the physical comfort requirements of the microenvironment of people's immediate office space are relatively constant over time and over building type.

Data interpretation has indicated that there are four critical components of spatial comfort to be addressed if the Spatial Comfort Crisis is to be averted. These are:

- the square footage of the individual workspace,
- space standards and practices,
- space trade-offs between the individual and the workgroup, and
- the degree and height of enclosure.

The ways in which space planners and corporate policies interact to resolve these issues dictate work-group layout and floor arrangements as well as furniture purchases and other large-scale monetary investments. Office size is discussed next, then space standards and practices. Issues affecting the allocation of space between individuals and group use follow; and partition height is discussed later on in this chapter, in relation to office noise control.

Office Size

Office size has traditionally been linked to rank within the organization, and still is. Employees expect bigger and more luxurious offices to be provided as part of their upward mobility through an organization. In companies where this is not the case, occupant feedback surveys generate furious comments because of fears that the results will be used to shrink office-size—and this regardless of physical configuration of the floor or the existing size of the offices. The space standards that many companies have developed to help them plan for future space needs as the organization expands and contracts vary somewhat, but not widely; what varies is the degree to which they are applied. In some companies, they are only a guideline, used on a discretionary basis; in others, they are a rigidly enforced code. Typical ranges are listed in Table 6.1.

Table 6.1   Typical Office Space Size Ranges

---

250 sq. ft. and up = senior managers, full enclosure.

100–150 sq. ft.   = management-level and professional employees, full enclosure.

40–100 sq. ft.   = workstations enclosed by partitions (heights listed below), for technical staff, administrative staff, and clerical support.

Low partitions (define the workstation but no noise protection or privacy) = about 40."
Medium height partitions (protect employees from noise, provide some privacy) = about 60."
Almost full-height partitions (simulate full enclosure) = 70" to 80."

Office size and height of enclosure together identify the rank and position of each worker, but not his or her function (tasks) in the organization. Space standards also specify how many worksurfaces, tables, file cabinets individuals may have, what type of chair, and whether or not they have a right to additional items such as bookshelves and credenzas. The introduction of systems furniture in many buildings, by standardizing space planning procedures and workstation components, has facilitated the application of space standards. Many employees find the dimensions of their workspaces shrinking as they are moved into systems furniture layouts.

In a large owner-occupied office building, a Building-In-Use Assessment was carried out before and after changes to the workstation design and layout of one floor. The building was relatively new at the time of the study, and the new systems furniture had been laid out according to the organization's space standards. Its success was important to the organization for future applications of the standards and future installations of the furniture. Building-In-Use Assessment was used because there were indications that occupants were unhappy with their space. On receiving the results of the first survey, the company initiated a detailed occupant participation process in which the precise nature of their discomfort was defined. Changes were eventually made to the layout of workstations to improve circulation, to provide more informal meeting spaces, to provide

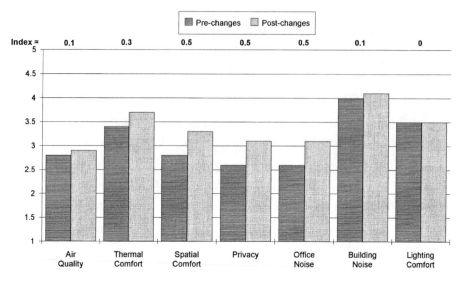

Figure 6.1. Comparison between BIU results before and after occupants participated in the replanning of their floor layout.

more definition of workgroup boundaries, and to provide support staff with more storage for workgroup files. In some exposed areas of the floor, higher partitions were added to increase people's privacy. Other than the partitions, these changes did not involve changing the furniture as such, or increasing the amount of space allocated, but as the results show, the impact of these changes was dramatic.

Figure 6.1 shows the comparison of BIU Assessments before and after changes were made to the floor. The results show that occupants' functional comfort increased significantly after the changes to their layout. Although all dimensions—except building noise control and lighting comfort—show improved ratings, the large gains have been in the three areas affected by furniture and spatial layout—namely, spatial comfort, privacy, and office noise control. A more detailed analysis of results shows that ratings from partitioned workstations located in the interior of the floor had in fact gone up so significantly that these occupants reported higher levels of functional comfort than the occupants of the enclosed offices with windows!

The BIU results helped the space planners determine what kinds of changes needed to be made to the other floors of the building—and indeed whether to make such changes at all. A BIU Assessment carried out later for the whole building showed that improved ratings occurred throughout, although these improvements were less pronounced elsewhere than for the prototype floor. This could be due to the involvement of occupants from this floor in a participatory planning process (negotiation with the designers) that was not used on other floors. The increase in the other floors' ratings, therefore, represent an increase in their functional comfort resulting from the changed layouts without the added benefit of environmental negotiation (Stages 4 and 5).

Space Standards

The ultimate space standard is the universal plan (UP). Also called universal footprint, this is a standardized layout provided to every workgroup and mostly fixed in place and not movable. It consists of a basic configuration of workstations, office equipment, and shared areas. The configuration can be adjusted as necessary to accommodate the requisite number of employees at different levels and with different tasks, but once constructed, is fixed in place. By investing in a UP system, companies feel they can control increasing demands for space as well as reduce the costs of churn.

Two Building-In-Use Assessments have examined universal footprint solutions. In one case, the space planners and design team wanted feedback from occupants of a universally planned floor to understand the strengths and weaknesses of the solution and to determine how to improve it for large-scale implementation throughout the company's buildings. The results of the survey showed that the new system suited some workgroups better than others. Fortunately, the sales and marketing employees that constituted the majority of the target group were those that felt most comfortable in the UP layout. Working in teams, they could see and hear each other as needed, but they had sufficient space to avoid hearing colleagues talk on the telephone, to meet informally, and to store their files. Those individuals that were least comfortable in the UP layout were support staff who had to store and access large numbers of personnel files, to discuss confidential matters on the telephone, and to function at the same time as an information resource for their group. For the support staff, the standardized openplan workstation layout was too open and too small. Eventually, this obliged the space planners to make a policy decision about the accommodation of secretaries with administrative responsibilities: in the interest of increasing the functional comfort of administrative staff, the planners opted to provide a larger and more enclosed individual workspace. This breakaway decision used functional comfort criteria to define more optimal accommodation, which meant overriding the traditional rank and status considerations that usually gave professional/technical staff larger workspaces than their support personnel.

In the second case, a company had moved into six floors of a new building. On five floors, employees had been moved in with their old furniture; the sixth floor had been planned with a newly purchased furniture system laid out on UP principles, and the space planning and facilities staff wanted to know how well employees were carrying out their tasks in the new environment on that floor. The workgroup accommodated in the UP layout was a professional group of architects and interior designers who had responsibility for office space design and planning for the large corporation in which they worked. Their work involved demanding visual conditions, required teamwork but also opportunities for individual concentration, and accommodated drawing boards in many of the cubicles. For this group, the partitions were too high to enable adequate interaction, and the lighting was not appropriate for the demanding visual tasks. However, the standardization of their work environment was accepted by occupants, and if the specifications for partition height had been developed more carefully and the installation adapted more closely to the floor to take advantage of ceiling light fixture and air diffuser placement, the UP installation would have been more effective. With minor adjustments, the planners were able to correct these problems and create a satisfactory UP environment of 80-square-foot cubicles accepted by professional staff who had previously considered themselves at a rank in the organization that required fully enclosed offices of at least 100 square feet.

Both of these examples illustrate the potential value of planning workspace according to functional comfort rather than square footage criteria. The end result in both cases was accommodating people in less space than they expected or thought they needed. In the first case, the sales staff performed well with less space than their support personnel; and in the second case, the professional and technical members of the space planning and design group performed well in 20 percent less space than they had originally occupied. As much of the emotional, territorial and irrational side of space-related issues develops around square footage allocations, using functional comfort criteria to make space decisions is advantageous in two ways—first, people are more likely to accept smaller dimensions for their individual work environment if other quality criteria pertaining to their workspace requirements are respected; and second, focusing on the quality rather than the quantity of space reduces the likelihood of emotional and territorial conflicts over office space.

The examples also show that using rank as the guide for space allocation is deeply embedded in corporate culture and only now beginning to yield to a more function-based approach to workspace definition. Not so many years ago, the AT&T headquarters building on Madison Avenue in New York City had two executive floors at the top of the building, which, apart from one or two lonely secretaries, were unoccupied; the executives actually worked out of other offices located on the premises of the companies they managed. In many companies, managers tend to use space as an informal system of incentive and reward. Some workers are still told "We'll have your space redone" to compensate when they do not get a raise; and some employers allow hierarchy to be gauged according to closeness to or distance from a window. This kind of reasoning, however, reinforces the territorial definition of space and, in the long run, generates more conflict than it solves. Workspace is not a form of currency: it is a tool. As in the case of the AT&T headquarters, managing space on a reward-for-status basis is not a cost-effective option for companies looking to increase the productivity of their employees; today, every space-related decision that is not considered on the basis of whether or not it helps the space function as a tool for work risks being wasteful of the company's resources.

Individual Versus Group Workspace

One of the reasons for the spatial comfort crisis is that even as planners try to decrease *individual* space size, the space needs of the *workgroup* are increasing. One area where this phenomenon is often experienced is relative to the common problem of the space required to store paper— files, drawings, records, archives. Whereas individual workers often insist on their need for file storage space that is both accessible to their workspace and secure from raiding by their colleagues, managers try to encourage their employees to reduce the amount of paper they store by returning files they are no longer working on and by using a centralized file storage and access system. In addition, many types of operation, such as those dealing with insurance or with legal responsibilities, are obliged to set aside large central areas for high shelving units to accommodate paper files, and many companies and government agencies fill warehouses in noncentral suburban areas with archival paper file storage.[2] Although many modern firms are investing in electronic file storage and retrieval systems, these systems are not yet the norm. For companies who are still debating the relative merits of a centralized filing system and whose employees keep large boxes in and around their desks, such as accounting firms whose archives of paper are legal documents, the leap into this kind of technology is only a distant possibility.

In spite of the advent of electronic technology, the paperless office widely discussed in the 80s is no longer a realistic vision. Adequate storage of work-related files is important for many types of employee, ranging from administrators with budgetary and personnel files to professional and technical workers who want project records and histories within easy access and preferably in their control. From a functional comfort viewpoint, files decentralized to individual desks are preferable to high shelves of centralized paper files as these tend to block the effectiveness of overhead lighting, impede airflow and generate uncomfortable levels of dust and particulates. In spite of the availability of new technology, relatively few companies have as yet made the investment in electronic file storage. Space planners fear with some justification that a lack of constraints on individual file storage in a paper-based system increases individual square footage requirements and renders less group workspace available for joint activities.

Other areas of increasing group space use include team meeting rooms and project rooms, as well as informal sitting areas and lounges, and space for shared equipment, such as faxes, printers, copiers, and specialized or dedicated computer workstations. The shift from the individual to the team as the focus of space planning solves the spatial comfort crisis by redefining space standards in terms of workgroup, rather than individual, space allocations which include, but are not limited to, individual and shared spaces. By defining the Universal Plan in terms of group rather than individual needs, individual workspaces become interchangeable; expensive and infrequently used equipment is not oversupplied, and the focus of the workspace is not the individual's desk but the shared space in which creative work is done. This shift in focus from the individual worker to the workgroup or team will not be accomplished simply through the redefinition of space and space standards, but derives from a carefully-planned values shift in the organization. It can be facilitated by appropriate environmental support for change. Providing the environment that encourages this shift, and indeed planning the environment that will accommodate such a shift, are both steps which require occupants' feedback and will only result from good communication and negotiation between managers and occupants of space.

THE PRIVACY CONUNDRUM

In the Building-In-Use Assessment system, privacy ratings are closely linked to spatial comfort ratings, and the overlap between the two in terms of physical office conditions is pronounced. For occupants and managers in North American office buildings, the concept of *privacy* is irrevocably linked to the notion of physical enclosure. And yet numerous studies have demonstrated that privacy as a condition of getting work done is more complex, usually extending beyond a simple physical element to the social organization and structure of the workgroup to which the individual belongs. Cultural, age, and job rank factors all affect definitions of privacy, which is largely concerned with the control a person feels over his or her accessibility to coworkers.

As the importance of the workgroup or team increases, one might expect the relevance of individual privacy to planning office interiors to decrease. However, for many types of task, some quiet space and time for individual concentration is often a necessary condition for their successful completion. These opportunities do not have to be provided in the form of an enclosed office or workstation. The challenge for managers is to determine how best to provide such opportunities for individuals in the context of work-group planning: it is clearly not a question of simply enclosing people, which can cause them to feel cut off. The conundrum posed by the concept of privacy is that the apparently conflicting need for social contact and joint work opportunities is at times as strong or stronger than the need for individuals to work alone, and that these needs vary according to task, work-group structure and individual personality differences. To resolve the conundrum, some kind of dynamic equilibrium between opportunities for solitude and opportunities for communication that can adjust to the needs of tasks being performed is the key to functionally comfortable privacy.

The Building-In-Use concept of privacy incorporates people's ratings of their visual privacy as well as their voice or conversation privacy. People also judge privacy by the amount of telephone privacy they have, both hearing other people's conversations and being heard themselves. The BIU norm for privacy is 2.3 in the first database, derived mostly from buildings with open plan floors containing not more than 30 percent enclosed offices, and the rest open-plan cubicles. In the second database, where the proportion of enclosed offices is closer to 50 percent, the norm for privacy rises to 2.9. This higher BIU norm reflects a greater number of enclosed office respondents. Respondents in open-plan cubicles consistently report lower privacy ratings than respondents in fully enclosed space, regardless of the height of their enclosing partitions. The concept of individual enclosure, then, is fundamental in considerations of privacy in most office workers' minds.

Interpreting privacy scores is a complex procedure that reflects the shifting balance of the Privacy Conundrum. High ratings mean more privacy, but high levels of privacy do not necessarily mean more functional comfort, because too much privacy means problems of isolation. Extreme privacy ratings both high and low are not easily interpretable. A high privacy score can indicate satisfaction with privacy, or it can mean the discomfort of feeling "cut off" from fellow workers and isolated. A low privacy score may indicate dissatisfaction with privacy, but work-tasks may not require solitude and concentration, therefore the low rating in this case does not mean a functional comfort problem. A lack of privacy does not necessarily impact negatively on all types of task because many tasks require a group effort or some social interaction or contact. The only way to judge the meaning of privacy ratings is to examine the specific task requirements of workers.

In summary, to understand the meaning of privacy in modern offices requires:

- analyzing the task requirements of members of the workgroup in terms of their need to concentrate (work alone) and their need to communicate (work together),
- determining the degree and height of enclosure of the individual workplace, and
- accepting that privacy is not an unqualifiedly desirable attribute of good office space: relatively few employees require one hundred per cent privacy, as most also require communication. Also, people are uncomfortable if they feel cut off from their coworkers.

People's privacy requirements are more complex than merely being able to enter an enclosed space and not be overseen or overheard, especially as team-based and project work becomes more prevalent. To plan an effective group workspace, the need for privacy may better be understood as a need for control over the work environment rather than the presence or absence of physical enclosure.

Evaluating Privacy

In the context of changing office tasks and workgroup organization, it is not a foregone conclusion that having less privacy is a bad thing. The strong division that still exists in many organizations between those in cubicles and those in enclosed offices has less to do with privacy than with the same rank-based arguments that are used to provide senior people with larger office sizes. Some organizations have tried to redress the imbalance by eliminating enclosed offices with windows and requiring managers to choose between an enclosed office with no window or a cubicle with a window. As space planners know all too well, providing enclosed offices next to windows effectively blocks natural light from reaching interior cubicles as well as views, and in those buildings in which supply air is delivered over or under the windows, ventilation as well.

Occupants of enclosed offices, although they assign high ratings to privacy, often confess that they do not fully need it, either because they are not there enough, or because they prefer to have contact with their workers, maintaining an open door and undraped windows. Many workers are happy to share enclosed offices with one or two or even three people. Paradoxically, even if they need quiet to concentrate, for many managers, their office is still not adequately private in that the telephone still rings, people drop by to talk, and there are frequent meetings. Most managers agree that they only close their office doors when dealing with individual employees on confidential matters such as evaluating their work performance or, sometimes, for telephone conversations on subjects not yet made public. Thus both for open-office and for closed-office occupants, privacy is less a matter of physical enclosure with walls and a door, and more a matter of environmental options. The concept of privacy is increasingly open to definition through management of space and time.

The floor plans illustrated in Figures 6.2 and 6.3 provide examples of the extremes that exist in terms of privacy in modern office buildings. The first floor plan (Fig. 6.2) illustrates a floor in one tower of the Great Arch at La Défense in Paris, France; it shows small, identical enclosed offices strung along long corridors. The floors in this tower are occupied by a large government department, which had moved from a series of low-rise war-time buildings where offices were grouped around circulation and meeting areas. When surveyed, the occupants of the Great Arch provided extremely high privacy ratings (3.4), but it was evident from their written comments that this was far from being a source of comfort. In fact, occupants of this building suffered from a feeling of isolation and a lack of social contact that seriously impeded their ability to do their work effectively.

In contrast, the second floor plan (Fig. 6.3) illustrates a fully open and mobile "free address" system implemented at the NAHB Research Center in Maryland, described in Chapter 2. Here employees have no designated workspace but sit at tables in a large open space, moving their telephones and their mobile filing cabinets as needed. The intention here was deliberately to reduce individual privacy to facilitate communication, teamwork and creative problem-solving. Whereas the Paris building occupants reported high privacy scores but in fact experienced

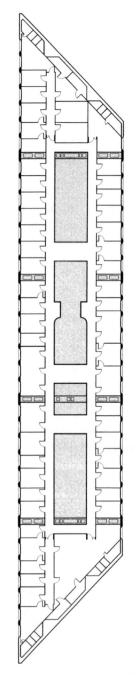

Figure 6.2. Floor plan of the Great Arch at La Défense in Paris, France.

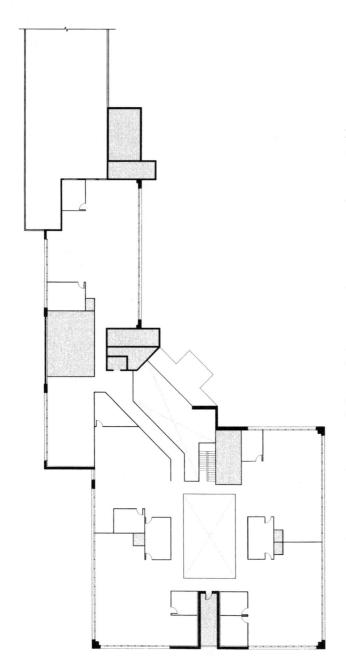

Figure 6.3. The second floor of the National Association of Home-Builders Research Center in Maryland, United States.

a lack of contact and communication that adversely affected the quality of their work, in the Maryland building, occupants had considerable trouble adapting to the new space and reported experiencing a disconcerting rootlessness resulting from a lack of personal "retreat space." Until they grew accustomed to the free address system, they actively withdrew from interpersonal communication and were temporarily at least unable to perform well in teams.

A compromise between total enclosure and total exposure is obviously a better answer than either of the above examples, and in many offices today a range of workspaces is being provided so that individuals and groups can move between them depending on the tasks they are working on and their workgroup requirements. Stone and Luchetti have developed a flexible approach to space planning which allows individuals not to be isolated in individual cubicles unless their tasks requires solitude for concentration.[3] According to their concept, the key elements of environmental choice as regards privacy include small group meeting spaces as well as a conference room, some enclosed offices for concentrated and private work, a selection of computerized workstations providing the software and file access needed by workers, and some small carrels or partitioned workstations where individuals can install themselves temporarily and store the items they need for work. As in solving the Spatial Comfort Crisis, the solution to the Privacy Conundrum is in "upside-down thinking" that all but does away with planned individual workspace and instead offers employees an array of spatial opportunities that range from individual to dyad to small and to large size groups.

THE CHALLENGE OF OFFICE NOISE CONTROL

Acoustic conditions have altered dramatically in modern offices because of the following changes:

- the large-scale introduction of electronic equipment
- smaller individual workspaces
- more meetings, group sessions and shared activities
- less physical enclosure
- increasing use of the telephone

The effect has been to increase overall sound levels in offices, making office noise control a more critical issue than in previous eras of office work. Physical interventions to manage noise better in offices, however, have been somewhat limited, making use primarily of different height partitions between individuals. As with building noise control, silence is not golden: most people prefer some level of background noise and activity to mask their own voices as well as to provide them with a sense of what is going around them. The Challenge of Office Noise Control is to maintain office noise at nonintrusive levels in a work environment increasingly favoring communication and collegiality.

Occupants' feedback on *office noise control* addresses the level of noise generated by coworkers, both through their activities and conversations and through the equipment that they use. Their keyboards, their telephones, and the drone of their disk drives all contribute to occupants' office noise control ratings as well as the general background noise of their conversations and the specific voices that can be heard with information-carrying content. Noise is not any sound: it is unwanted sound. It is therefore by definition uncomfortable and stressful in the work environment, but not necessarily threatening to occupants' health and well-being.

The BIU concept of office noise control is based on people's ratings of general noise levels, of specific intrusive noises, such as others talking or the operation of nearby equipment, and the degree of disturbance caused by the noise. The BIU norm for office noise control is 2.9 in the first database, and 3.0 in the second, meaning there is no real difference in office noise between older office buildings and the buildings in the newer database, which are more densely laid out even though they have more enclosure. Most feedback shows a strong link between spatial comfort, privacy, and office noise control scores. Types of work that require a high level of individual concentration, or sustained telephone conversations, or which have confidentiality requirements, seem to be the types of task most impacted by office noise levels. What are the most common sources of office noise problems?

A growing proportion of office work is carried out on the telephone. Employees who are involved in sales and marketing, in customer service, in providing information, in account management and billing, and in many other areas of direct interface with customers spend much of the day on the telephone. Although their work does not have confidentiality requirements, groups such as these are affected by noise. In an effort to manage noise better for one such group carrying out telemarketing, people were moved out of a completely open "bull-pen" layout into systems furniture that enclosed each person in 60 inch partitions. One of the aims of the change was to enhance acoustic conditions for telephone-based work, but the new layout also reduced the amount of space per person on the floor. The results showed that although employees' heard less of what their colleagues were saying after the move, they became increasingly uncomfortable about being overheard by their neighbors because they were closer to them. They expressed fears that their colleagues would listen to and criticize their marketing approach. Moreover, having been provided with space that could be personalized, workers pinned up pictures, calendars, and posters on the inside of their partitions, effectively reducing the sound-absorbing surface inside their cubicle; as a result, the overall noise level of the group actually increased.

Other sources of noise are equipment (fax machines, copiers, printers, and computer keyboards), movement (people moving around, going in an out of meetings, and talking to each other while walking), and simply people talking to each other or on the telephone. A Building-In-Use Assessment was carried out on occupants of a newly completed office building who complained that they could not work because too much sound was being transmitted through the open-plan layout, and that they were disturbed by hearing each other so clearly. These individuals had moved from an older building where they had had bigger individual workspaces on larger, darker floors. Thus the lower partitions of the new furniture gave them access to window views and natural light, but caused them to lose the voice privacy they had previously enjoyed. Their ratings compared to BIU norms are displayed in Figure 6.4. As ratings in this building are otherwise positive, it is all the more surprising to note occupants' low ratings of office noise control. This result illustrates the importance of workers' previous experiences in affecting their judgment of acoustic conditions. Unlike the other functional comfort dimensions, noise control ratings are unduly influenced by previous environmental experiences and respond well to familiarity and behavioral adjustment over time.

Office noise only becomes a problem if it exceeds a certain level of intrusiveness, usually by being irregular and too loud. This is less related to measurable sound levels than to the content and meaning of the sound that can be heard. For example, being able to follow every word of someone's conversation is far more disruptive than an equally loud distant conversational buzz in which no words can be made out. A large number of office noise problems are related to use of the speakerphone. Whether they are in cubicles or in offices with the door open (and sometimes if it is closed) speakerphone conversations can be a major source of noise pollution in the workspace, often carrying through full-height partitions as well.

Other types of noise can be intrusive because of their meaning content. In one building where office noise was a problem, technicians and draftsmen seated in open-plan workstations objected to being able to hear their coworkers' radios and fingernail clippers! These sounds

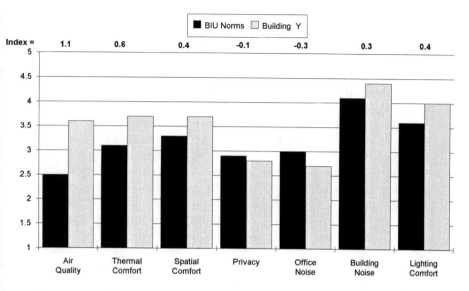

Figure 6.4. Building-In-Use profile of Building Y, showing office noise discomfort in an otherwise comfortable building.

were not loud, they were objectionable because they were meaningful—they carried information. In such cases, taking sound pressure readings with instrumentation is not likely to yield evidence of noise discomfort, but the occupancy feedback informs us that it is there. Instrument measures of noise, such as sound pressure levels, sound reverberation, and sound transmission levels, are not adequate to predict people's acoustic comfort in the workplace. Stress from noise discomfort can adversely affect the speed and accuracy with which employees perform their tasks, can increase fatigue, and can have a serious impact on functional comfort.

Noise Control Through Partition Height

People need to hear each other work because they do not want to work in isolation. In a quiet environment, individual voices and other information-carrying sounds are transmitted throughout a space. Some degree of background sound is therefore desirable, and some groups tolerate relatively high levels of sound, whereas noise—unwanted sound—can also easily reach intrusive levels that affect the efficient performance of work. Functional comfort relative to noise control can be effected by increasing distance between workers and noise sources (i.e., equipment), by spreading people out, or by enclosing and shielding noise-generating equipment or individuals. Many managers select some kind of partition system for individual workers in an effort to manage office noise, with varying success.

Workstation partitions are used in part to control sound transmission in an open plan office, and in part to protect individuals' voice privacy, as well as to provide visual screening. Some workers with none or with low partitions find this acceptable because of the work they do. For example, receptionists need to see and be seen in order to do their job, and they have less reason to protect themselves from hearing or being heard by others than other types of worker. Data entry clerks or other workers at terminals who are all doing the same kind of work rely on contact with co-workers for effective completion of collective tasks, as in the case of the "Just-In-Time" configuration developed by the Bank of Boston and described in Chapter 3.[4] Many types of customer service representatives working on telephones need to see coworkers and to be seen by supervisors, and they accept the sound of each other's voices as part of their shared work-group culture.

Employees with medium height partitions (48 inches to 65 inches) are more acoustically protected than those with lower partitions, often performing tasks that require more individual concentration and less interaction with colleagues. Although research shows that a height of about 60 inches is optimal for sound absorption as well as voice privacy between workstations, the sound of coworkers' voices still travels over and, if there is space, under the acoustic partitions that separate them.[5] As in the example described above, it is not uncommon for the acoustic properties of partitions to be rendered ineffectual by personalization and decoration. Occupants' posters and photographs reduce the expanse of sound-absorbing material and increase the ratio of sound-reflecting surfaces.

In some offices, partitions of six feet or over are used in lieu of enclosed offices. Occupants' feedback repeatedly demonstrates that people are more uncomfortable inside high partitions than inside low and medium height partitions. There is an ambiguity to being accommodated in an all-but-enclosed space while still being able to hear one's neighbors and know that one can be heard. Occupants' functional comfort in terms of both noise and privacy is therefore not increased by high partitions. It is more cost-effective for noise control purposes to accommodate workers either within low or middle height partitions or in fully enclosed offices. This is not to say that acoustic conditions are necessarily superior in enclosed offices: the acoustic insulation of full-height partitions or walls depends on the standards of construction used. The more acoustically effective the construction, the more expensive it is to build. Some fully enclosed offices are constructed with full-height demountable partitions which provide no acoustic insulation at all. In some buildings, walls are built to extend above the ceiling, but holes are cut in them for radiators or ventilation units that straddle two offices. These conduct sound very effectively, negating the insulating value of building walls up to the slab!

The relationship between partition height and noise control is not as simple as it appears. As with all the elements of the interactive user-environment system, it is more complex than simple cause–effect thinking would have us believe. Whereas occupants believe there is a linear relationship between height of partition and acoustic effectiveness—the higher the partition, the greater its effectiveness in controlling noise—in fact, intervening factors such as type of wall construction and ceiling finish, size of workspace (and therefore distance between people), distance to windows, amount and type of equipment (including speaker-phones) and type of work (telemarketing or personnel counseling) all have to be addressed in order to make good noise control decisions.

Just as the previous section showed how privacy is associated in people's minds with physical enclosure—even though functionally comfortable privacy is more complex than a single physical element—similarly office noise control is usually associated with partition height. Yet as we have seen, as a functional comfort element, the impact of noise is in fact somewhat independent of partition height. Both these functional comfort concepts are far more complex than one simple physical element, but nevertheless these simple physical elements (walls, partitions, doors) have a symbolic role which is clearly important to occupants. In developing responsible accommodation strategy, the wise manager recognizes these discrepancies between the symbolic significance and the functional comfort of elements of the work environment. It is all the more important, therefore, that strategic decisions not be made without communicating with users about their space.

THE FUTURE OF THE INDIVIDUAL WORKPLACE

Key themes have emerged in the planning and design of office interiors that indicate the need for some upside-down thinking to resolve space-related dilemmas and lead the way toward the future.

1. The standardization of space requirements to ensure cost-effective allocations of office space must be based on a logic of group task requirements and functional needs rather than on a logic of individual rank and status, or even individual needs.

For most North American companies, individual space standards are going down while the overall amount of space per individual in a building is going up. As the costs of acquiring and of operating space are also going up, it is imperative to determine the most cost-effective ways of using and planning work space. Meeting individual needs for privacy, status rewards, and personal space is increasingly unrealistic for companies seeking to add value through an accommodation strategy. On the other hand, defining and meeting the needs of the business unit, or the work-groups within it, through a systematic approach to feedback, task analysis, and environmental negotiation is a very promising approach to streamlining costs and increasing value.

2. Knowledge work requires opportunities for individual focus and concentration as well as teamwork. Much repetitive computer-based work requires sustained worker contact, long hours of immobilization.

Workers not tied to computer terminals spend time on a range of activities with different requirements that are not available in a single workspace. Many office workers therefore have less need for an individually-designated, personalized workspace, and a growing need for a choice and variety of spaces to suit their tasks. On the other hand, as we saw in Chapter 2, modern offices also accommodate the terminal-based worker who spends his or her working day on a computer, often accessing other equipment at the same time, such as microfiche files, and/or dealing with customers on the telephone. Whether these workers are temporary and replaceable, or part of the permanent work force, they have stringent requirements from their work environment on which a significant proportion of their productivity depends. They are virtually locked into their workspace, so the right lighting, sound level, thermal comfort, and furniture design are key ingredients in the performance of their work. These two extremes in workspace requirements both need to be accommodated in modern office space.

3. The workgroup is a more effective unit for workspace planning than the individual worker, especially in terms of work storage and file access. Shared workspace considerations override the traditional concept of the individual office.

The individual workstation or office will wane in importance as new and innovative ways to work become more prevalent in organizations. As people are moved out of office buildings into home-work, satellite offices, or mobile workstations of some kind, the space remaining as home base for employees will be increasingly shared and increasingly social. The spatial comfort crisis, the privacy conundrum and the office noise control challenge will be solved when almost no individual workspace is assigned on a permanent basis, and most offices are group territory and a tool for the group's work.

In conclusion, management of noise, privacy, and spatial layout is an increasing challenge for changing corporate needs. Computer-based workers tend to tolerate more office noise and poor privacy than other groups, but this is not to say that they could not work better in more functionally comfortable workspace. Knowledge workers tend not to sit at a desk, but move around the building and seek out opportunities to interact with coworkers. More and more workers in the future will spend time working outside the building. Future office interiors must accommodate all these different needs, and they cannot do this by performing yet another variation on the old theme of the individual office space filled with heat-generating equipment, located by the window for managers, with a conference room available on another floor. Planning environmentally for these changes will be successful if functional comfort criteria are defined and respected in workspace design, if workers are encouraged to understand and participate in planning their workspace, and if the communication of information about the user-building relationship is encouraged.

In summary, the spatial comfort crisis is resolvable through space planning and standard setting that addresses work-group rather than individual worker needs; the privacy conundrum becomes understandable if privacy is defined in terms of space and time opportunities needed for individual tasks rather than of enclosure and isolation; and the office noise control challenge is effectively met through creative reexamination of the function and performance of partitions as well as balancing team-space requirements with individual needs. In the next chapter, the final Building-In-Use dilemma is presented as an opportunity. Lighting comfort represents a significant opportunity to business managers because of the range of innovative lighting technology that is available. Lighting comfort, a critically important aspect of users functional comfort, is connected both to building systems and to the planning and design of interior space, yet stands on its own as a key determinant of the speed and accuracy of work performance.

NOTES AND REFERENCES

1. See Semler Richard "Managing Without Managers" *Harvard Business Review*, Sept.-Oct., 1989; p. 56 and Condon, Ron "Goodbye to all this" *The Times* of London; 26 May, 1995; p. 36.
2. A legal ruling in August 1993 obliges the US government to print out all electronic communications on paper in order to maintain an accessible record.
3. Stone and Lucchetti, "Your Office Is Where You Are," p, 102–117.
4. See Chapter 2 for a discussion of Just-In-Time office design. Sraeel "Bank of Boston's JIT Gives Eileen Harvard the FM Edge,"p. 410.
5. M.David Egan, *Architectural Acoustics*, (New York; McGraw-Hill Book Company, 1988,). p. 17.

CHAPTER 7

BUILDING-IN-USE ASSESSMENT OF LIGHTING COMFORT

"Lack of daylight is a factor that gets on our nerves. You feel as though you're in a cage. There are not enough windows. A happy employee, satisfied with his work environment, will show it through his work."

building occupant

THE LIGHTING COMFORT OPPORTUNITY

Since the widespread introduction of computer screens into modern offices, lighting the work environment has become more complex and difficult than it ever was in the factory. Researchers are learning more about the complex human response to light; lighting technology for buildings is evolving rapidly; yet, in the 1980s, eyestrain and sore eyes were the most frequently reported health problem in office buildings.[1] The opportunity for the 90s and beyond is to find ways of installing better and more responsive lighting in the work environment and watch productivity soar.

Compared to ventilation and temperature conditions, people in offices complain relatively little about lighting and appear to accept and adapt to a far wider range of visual conditions than temperature and ventilation conditions. People's expectations that their comfort needs will be met seem to be more developed with regard to air quality and thermal comfort than to lighting comfort, so they are more demanding regarding ventilation and temperature than they are regarding glare, gloom, contrast conditions, and light levels. The human eye is a highly adaptive mechanism, and it would seem that people are used to making it work under what can seem like uncomfortable visual conditions for long periods of time before sensing discomfort and demanding improvement.

Of all the functional comfort dimensions, *lighting comfort* shows consistently the greatest discrepancy between human visual comfort requirements and the quality of lighting supplied in the workplace. Business managers concerned with the O–A relationship therefore stand to make the most substantial gains from investing in this functional comfort dimension. The Lighting Comfort Opportunity is significant in the following terms:

- Lighting quality is obviously related to the performance of visual tasks, and 70–80 percent of all the tasks performed in an office are visual tasks. Improved lighting quality in the workplace improves task performance and reduces eyestrain, so productivity should clearly improve.

- Energy-efficient lighting is a major cost-saver in modern buildings. Energy-saving programs through subsidized lighting technology are now available in most states and provinces in North America.

- Existing lighting technology offers opportunities for a degree of individual control over lighting that is as yet a remote possibility in areas of ventilation and thermal comfort. Investing in lighting controls increases functional comfort by enabling individual differences in visual requirements to be accommodated easily.[2]

The problem—or rather the opportunity—of lighting comfort, therefore, is to invest in lighting technology that can effectively respond to the delicate balance between the human eye's ability to adapt to a wide range of conditions, and the varied and even conflicting requirements of the modern office environment. Until the large-scale introduction of CRT (computer) screens into the office work environment, designers specified office lighting to be as uniform as possible and bright enough to permit close visual tasks such as reading and writing to be carried out at every work surface. This approach was based on lighting requirements in manufacturing environments where similar, standardized tasks were being carried out and uniformly high light levels seemed appropriate. However, as office work has diversified and visual tasks are more varied and demanding, appropriate lighting for work has become a far more complex issue. Visual tasks in modern offices include computer work on a wide variety of screens, meetings with small and large groups of people, reading and writing, one-on-one work sessions, drawing and design, reading and, in some cases, annotating computer printout, and work with microfiche, video, and a wide range of graphic displays.

In spite of this rich array of needs, lighting *performance* is still mostly measured in terms of amount of light emitted by fixtures, *Illuminance*, and lighting *effectiveness* is mostly assessed in terms of amount of light at the work surface, *Luminance*. Efforts to incorporate some of the other factors affecting visual comfort—such as contrast conditions and glare, color differences, and modeling or directionality of light—into a scale of environmental measurement are less widely used because of the complex difficulties of measuring them.[3] Comfort specifications for lighting are usually limited to recommended light levels, as measured in footcandles or lux (metric), and, in recent years, the amounts considered appropriate for office work have been reduced to reflect the widespread use of CRT screens.[4] On the other hand, a familiar sight in most offices are ceiling fixtures with one or both lamps extinguished by employees who want less light on their computer screens, and even on their work surfaces. Many office workers bring in their own incandescent desk lamps.

From the occupants' point of view, access to natural light is at least as important as good quality artificial light, and perhaps more, even though windows may cause glare conditions that are adverse to work at computer screens. In spite of a surge of interest in daylighting design a few years ago, mainly for energy conservation reasons, the need for people to have access to natural light seems to be less important in North America than it does in Europe. Office workers themselves are often aware that it is against the labor code in certain European countries to accommodate office workers in windowless rooms. Yet, in North America, it is not uncommon for people to work in basement offices and in rooms and cubicles in the interiors of deep floors. Proximity to the window has come to be considered a sign of rank rather than a human right, and as a result in traditional organizations the more highly-paid professionals and executives have offices next to windows while lower-ranking staff are usually located in interior offices and cubicles.

From a functional comfort viewpoint, however, whether the light comes from windows or from light fixtures is less critical than the effectiveness of lighting in facilitating people's tasks. From the point of view of personal health, people gain more from going outside a building at lunch or on breaks than from sitting inside looking out through a window all day long.[5] Designers know that while increasing daylighting in the workspace may increase people's *psychological* comfort, it can produce screen glare from windows for people working on computers. Similarly, lower lighting levels in offices may improve screen-based working conditions, but can generate gloomy, underlit work environments that lower morale and cause complaints.

The Lighting Comfort Opportunity is therefore a complex one, comprising elements of electric lighting technology, daylighting design, complex visual task analysis, visual differences among workers, the psychological effects of light, and the relationship between space, furniture, color, and amount of light. People have an apparently conflicting need for, on the one hand, lowered background (surround) light levels to reduce glare problems on screens, and on the other, for a bright, colorful, and luminous environment to keep up morale and enhance the aesthetic dimension of the work environment. All people, however, benefit from bright lighting of their visual task. They have a need to vary light quality and amount depending on the requirements of the work they are doing and the environment they are doing it in. The lighting comfort opportunity is therefore a major one whose potential has largely been overlooked in workspace design to date.

BUILDING-IN-USE ASSESSMENT
OF LIGHTING COMFORT

Analysis of occupant feedback on window access and proximity confirms that whereas the natural light aspects of lighting comfort contribute to occupant morale, it is the conditions created by the electrical lighting system that have the greatest effect on occupant health, probably in relation to eyestrain. The BIU norm for lighting is 3.3 in the first BIU database and 3.5 in the second. The first database comprises buildings whose lighting predates the large-scale installation of computer screens. The ceiling fixtures in almost all these buildings are recessed, two-bulb fixtures with acrylic lenses designed to disperse the light over a wide area and minimize dark or low-light areas in the visual field (see Figure 7.1). As a result, most of the complaints received about lighting in such buildings pertain to glare and discomfort from overly bright fixtures.[6] In many Canadian office buildings, light fixtures are installed in a coffered ceiling system on the grounds that recessing the fixture means glare is reduced. This may be true when one is standing up, but from the seated position in which most people work, all fixtures are in full, glare-generating, view, similar to a conventional, flat ceiling. An example is shown in Figure 7.2.

coffered ceiling -- recessed fixtures   *directional downlighting -- parabolics*

Figure 7.1. Difference in direction of light emitted from a coffered ceiling and from louvered parabolic light fixtures.

Figure 7.2. Typical coffered ceiling configuration, with two-bulb fixtures behind acrylic lenses.

The second and more recent BIU database, although it shows a higher (more comfortable) norm for lighting comfort, includes a much greater proportion of VDT workstations as well as a wider variety of lighting systems. The second BIU norm may reflect the improved management of glare in newer buildings. In United States office buildings, glare is partly reduced by installing directional lenses on ceiling-mounted fixtures so that the light is not diffused but illuminates only the work areas immediately beneath where tasks are being carried out (see Figure 7.1).

The worst rated building for lighting in either Building-In-Use database is not a North American building at all but a new office building

Figure 7.3. Office in the Great Arch at La Défense (in Paris) showing the large window and dropped ceiling light fixture.

in Paris, France (See the floorplan in Figure 6.2). Offices in the Parisian building are enclosed, accommodating one, two, or three people, and are all located along the perimeter walls of this long, thin building. All have at least one large (60 inches square) window occupying most of one wall. There is a space of some 18 inches between the interior glass and the exterior glazing. Roll-down blinds intended to reduce light but not views are inadequate for the amount of light on the higher stories. The building is 35 floors high, and above floor 6 there are no adjacent buildings to cut the bright light from the sky. In addition to the large bright window, there is a single fluorescent fixture double the length of the average North American fluorescent dropped down on two rods at right angles to the window (see Figure 7.3).

The voice, power, and data outlets are located under the window, with the result that most people with PCs on their desks face the window. Each desk—the main work surface in each office—is placed under the hanging fixture. These lamps are only switchable in banks of 6–8 offices, so out of respect to their colleagues, occupants leave them on whether they need them or not. As a result, all the visual tasks that the occupant might conceivably be performing in his or her office are carried out in a glare situation from both the window and from the fluorescent lights. The lighting comfort score received from occupants of this building was 2.8 on the 5-point scale, where the norm is 3.3, an unspeakably low score by North American standards, where office workers rarely rate lighting comfort lower than 3.

Lighting comfort is difficult to measure because occupants are often unaware of visual discomfort until it is too late and eyestrain has occurred. They are also likely to attribute the physical symptoms of visual discomfort such as headaches, dizziness, and fatigue, to indoor air quality problems of which they are more aware. The lighting comfort problem comes from the fact that while visual tasks in the office have diversified enormously in recent years, office lighting environments generally have not. The Lighting Comfort Opportunity is that the technology is available to enable environments to respond more appropriately to occupants' functional comfort needs as soon as managers incorporate functional comfort criteria into their accommodation strategy.

HUMAN FACTORS IN LIGHTING

Given the exquisite delicacy of the human visual mechanism, it is not surprising that assessing lighting comfort is more complex and nuanced than assessing other functional comfort dimensions. Whereas workers simply provide low comfort ratings for thermal comfort and office noise if they feel impinged upon by these conditions in the performance of their tasks, building occupants do not necessarily rate lighting comfort as low: they simply do not rate it very high. Part of the difficulty inherent in assessing lighting comfort is occupants' own predilection to accept adverse functional comfort conditions in lighting more readily than they would in other areas of functional comfort.

Studies of visual comfort at work recommend lowering overhead light levels, and installing combination task-ambient systems that light only the visual task; they also advocate strategies to manage glare and reflections.[7] At least one study has suggested that the incidence of headaches and eyestrain could be halved with the use of high-frequency (controlled by ballasts) lighting installations, but this type of information, like many of the recommendations from other lighting studies, has not yet been systematically applied to office lighting design.[8] These authors conclude that occupants' reports of headaches and fatigue are likely to be linked to presumed air quality problems by both occupants and managers. The trend in recent years, however, has been towards reducing the amount of overhead light to accommodate screen-based work as well as to reduce high energy costs by providing lower light levels in areas where people meet, circulate, converse, and perform other tasks that do not require close visual work. As a result, *Illuminance* standards for overhead light levels have been going down, and task-ambient solutions with supplementary light provided at each desk is becoming a more common situation.

In many modern office buildings, fixtures with flat, acrylic lenses that diffuse light widely and evenly to either side have been replaced by parabolic fixtures with "eggcrate" diffusers that do not disperse light the way the flat acrylic lenses do. The parabolics cause the light to focus directionally, lighting what is directly underneath them and leaving the space around somewhat less lit. The eggcrate or louvered diffusers allow the light to be projected downwards, thereby protecting occupants seated elsewhere from glare from these fixtures (See Figure 7.1).

A recent study compared occupants' comfort and satisfaction in offices with parabolic downlighting (directional) and in offices with lensed indirect (uplighting) systems and found that people report better levels of comfort, productivity, and satisfaction in the uplighting environment in which the overhead fixtures do not project light down at all, but provide an even, background wash by reflecting light up, onto the ceiling and walls.[9] Indirect lighting systems such as this one ensure that each workplace is provided with task lighting that projects an appropriate amount of light on to the visual task (not the CRT screen). In spite of this finding, the louvered parabolic has become the fixture of choice in most North American office buildings in response to the proliferation of screen-based work. This design decision may be in part because it is less expensive to install than indirect lighting, and indirect lighting requires higher ceilings than are available in most office buildings.

Parabolic (directional) downlighting is not inappropriate in most office layouts, but it presents certain problems. The spaces between light sources are often perceived as quite dark by occupants, and the spacing of workstations has to correspond precisely to the location of ceiling fixtures to take full advantage of the light. Slight moves and changes in space layouts over time, or the effects of "systems creep" when a new furniture layout is installed and work surfaces are no longer in lighted areas, can very quickly produce a less than functionally comfortable lighting environment. Also, people seem to prefer a bright ceiling—perhaps because it feels like the sun is above their heads—and directional fixtures, by throwing light straight down, darkens the ceiling.

An indirect lighting system can provide individual workers with more control over their immediate lighting environment through task light controls. Lighting controls increase functional comfort levels and can therefore provide a measurable payback in terms of increased speed and accuracy of worker performance. The quality of the task lighting, however, is critical to the success of such an innovation. A recent study showed that office workers working under the low levels of background lighting provided by indirect uplighting, and who had bright but unadjustable task lights at their desks, reported the lowest levels of both satisfaction and productivity compared to people working in other lighting systems.[10] These reactions were in spite of the fact that the actual amount of light for their visual task was the highest in the sample. The study considered the problem to be too much light. The issue of individual control, therefore, would seem to be more critical than measurable amount of light in responding to the complexity of people's lighting comfort in the workplace. A variety of individual control mechanisms are currently available.

LIGHTING AND MORALE

New lighting standards that recommend lower light levels in order to accommodate screen-based work do not solve the related functional comfort problem: that of worker morale. Results from Building-In-Use Assessments show repeatedly that lowlit and underlit space, even if it comprises circulation and shared areas (for example, where copy machines are placed), create gloom, in the sense of poor visual discrimination in the periphery of the visual field, and a condition known as "poor color rendition." These factors both affect building users negatively.

The effect of working in what are perceived as gloomy surroundings lowers occupant morale. Data interpretation of low lighting comfort scores indicates that people dislike darkened corridors and the darkened walls and ceilings that directional ceiling lights cause. People who sit in interior offices without windows often require major increments in light levels to perform the same tasks as workers in cubicles or offices with windows. Although the effects of reduced light in the visual field cannot be said to impede work performance for tasks carried out at the workstation, occupants report discomfort if they perceive their other

workspaces as gloomy. For example, badly lighted meeting and conference rooms can impede effectiveness when people shorten meetings
and look for other spaces, even though reading and writing tasks are
minimal in these spaces. Low lighting in corridors make distances seem
longer. And in areas where people have turned off their own overhead
fixtures because they are working on VDTs and are trying to reduce
glare, a real danger exists from loose cables, stacked boxes, and other
floor hazards that are less visible than they should be.

What is needed in these situations is not simply more light, because
this can generate glare as well as increase energy consumption. What is
needed is lighting that brightens and cheers and improves color rendition, without necessarily increasing *Illuminance* to the levels usually
specified for work surfaces. Studies have shown that good color rendition can counteract bad lighting design, and that using appropriate finishes and the right kind of lamp means that good color rendition
compensates for overall less light.[11] Workers in interior, windowless offices have expressed a preference for full-spectrum lamps which, even
though they are fluorescent, provide more daylight-like illumination. In
one building, employees went so far as to use their office supplies
budget to purchase some of these lamps and install them themselves
because building managers claimed they were too expensive and would
not result in tangible improvement. They claimed that there are no
well-controlled studies unequivocally establishing the benefits of full-
spectrum lamps, and that other choices are available to meet the functional comfort requirements of workers who want a bright visual environment without glare. Using innovative lighting design, installing
triphosphor lamps for better color rendition, and placing fixtures pe-
ripherally to reduce dark shadows can create a bright, quasi-daylit environment which does not rely on task lighting—which may be
insufficient—but which does not create glare on screens.[12] Also available is a variety of daylighting technology, some as unusual as light
pipes, which can bring daylight without glare, and even with views,
into every interior space.

The pronounced aversion building occupants express for apparently
underlit, gloomy environments, even to the point of preferring dysfunctional levels of brightness, may be related to Seasonal Affective Disor-
der, the emotional/physical reaction which is caused by reduced
daylight as the seasons change.[13] Even if this effect does not attain path-
ological levels in the majority of the population, there could well be a
physiological basis for preferring bright to low light levels in the work
environment, which is, after all, where adults between the ages of
twenty and fifty spend the majority of their time.

WINDOWS AND DAYLIGHTING

Better comfort ratings are generally received from building occupants with access to windows than from occupants seated away from windows in virtually all types of building. Some companies have made an attempt to equalize this situation. For example, one company (mentioned in Chapter 6) offers its managers the choice between a windowless enclosed office and a cubicle by the window so that no enclosed offices are placed by the windows and natural light is admitted as far as possible into the floor. In spite of evidence from occupant feedback that managers who have chosen cubicles rather than enclosed offices in order to have access to daylight complain about their acoustic privacy and their exposure to office noise, this policy has been accepted throughout the company's office accommodations.

Other companies trying to improve daylight accessibility have fitted the interior walls of each enclosed, windowed office with clerestory panels, so that daylight, if not a view, is transmitted into the interior of the floor. This design policy, although it increases the unit cost of interior wall construction, has yielded an improvement in the lighting comfort scores of occupants seated away from windows. Lighting measurements, however, do not indicate that people seated away from windows in these floor layouts receive enough daylight to illuminate their visual tasks. A spate of buildings in the 1980s experimented with various types of daylighting design to bring natural light into the interiors of office floors. These efforts, however, were more to control the extensive energy consumption of artificial lighting systems (30-50 percent of an office building's energy budget is consumed by electric lighting) than a response to the human need for natural light. Since the energy crisis has passed, efforts in this domain have dropped off, but they will no doubt return as energy issues regain their importance.

The window, with or without daylighting design, remains, in people's own opinions, one of the keys to their functional comfort in office buildings. Windows are important because they supply light, air, and view to the work environment. Although it is difficult to think of windows as supplying any one of these without the other two, the solution to the problems of windows in offices—who should sit next to them. Should they open? Should people be seated where they cannot see them?—is to consider the three different phenomena associated with them as three separate items. The ventilation aspect of windows is related to air quality problems; the daylight aspect of windows is related to lighting comfort; and the view aspect of windows is related to employee morale. The Table 7.1 summarizes the key differences among

Table 7.1   Design Implications of Different Window Functions

Windows as Ventilation	Windows as Light Source	Windows as Providing View
Windows open to the outside to permit fresh air to enter. They do not work as well if they are too high above the ground, and in urban areas opening them can result in excessive noise coming in from outside.	Windows are glazed openings in the exterior wall that permit light to enter. They can also permit heat to enter. Light from windows should reach as many people as possible, but the heat should be controlled.	Windows are glazed openings directed very specifically towards a visual experience which contains color, movement, and variety. Such openings do not need to be on an outside wall. They simply permit people on one side to look at what is happening on the other.
Design Implications	**Design Implications**	**Design Implications**
Design windows to admit small amounts of air when opened and to be quite sealed when closed. This specification is less feasible in high buildings, but the windows do not have to have large apertures to be effective. Operable windows have a critical positive impact on employees' perception of indoor air quality, and the sophistication of modern HVAC technology is such that adequate ventilation conditions can be maintained for the supply and circulation of fresh air through the mechanical systems even when air is infiltrating through window apertures.	Windows are large and often horizontally seamless, designed to admit maximum light but usually with glazing that filters out some of the light and resists heat. Smaller apertures are usually considered more energy-efficient. Some protection against large expanses of sky is usually required to protect occupants from glare, which makes roller blinds or horizontal louvers more effective than the vertical blinds that are more common. Clerestories work well to bring daylight into interior spaces.	View windows do not come with glazing or size specifications, but are likely to be more successful if the view is active rather than of a blank wall or roof. Interesting views should be provided without intruding on other people's privacy. Many atria, for example, are enclosed by windowed offices that admit light to interiors but which also permit direct visual access into each other's workspaces. This sense of being overlooked and watched throughout the day intrudes on people's functional comfort: it can impact their effectiveness at work.

windows defined as ventilation, windows as lighting, and windows as view-providers. In each case, design implications are summarized.

These three window definitions have important design implications, which are not necessarily the same or even compatible. For example, the *window as ventilation* lets air in, also admitting noise and possibly odors. Its positioning in the facade of the building, and its orientation are, therefore, important. The *window as light source* also allows heat gain to occur, especially where there is sun exposure. Moreover, because the most amount of light is admitted when the glazing admits light reflected from the sky, and light from the sky generates glare, this option can be problematic for people working at computer screens. Thus the glazing in these windows should be heat resistant, and preferably filter out some parts of the light spectrum. The *window as view-provider* has to have a view worth looking at, but views into an atrium can and do have an impact on the privacy of other workers.

Just providing lots of windows, however, is not an effective or thoughtful response to the human need for daylight and for good window design. A building designed specifically to admit maximum daylighting for occupants (as well as to be energy-efficient) generated the Building-In-Use Profile shown in Figure 7.4. The very low lighting comfort score resulted from the extensive glazing which created an uncomfortable glare situation, especially for people on computer screens. The fact that almost everyone in the building worked at computer terminals had not been taken into consideration when the building was being designed in the early eighties. The building has nine mini-towers (of about 6 stories), each of which is octagonal and glazed on all sides,

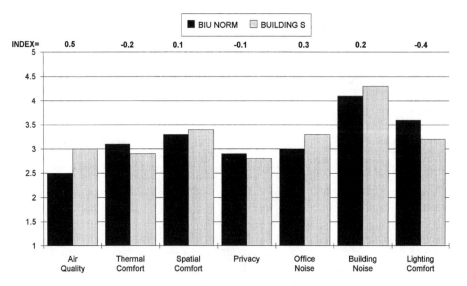

Figure 7.4. Building-In-Use profile of office building designed for natural light, where occupants suffered from heat gain and glare.

grouped around a large, glazed atrium. Daylight is admitted through the numerous windows as well as reflected off the windows of the other towers. Although people have a tendency to think that, like privacy, the need for natural light is limitless, in fact, as with privacy, it is not. There is a point at which there can be too much natural light for employees to work effectively, and at which the presence of windows means too much heat when the sun is out. In this building, a range of different window screening techniques was being tested to determine how best to control both heat gain and glare: an expensive retrofit that will be problematic over the lifetime of the building.

The three functions of windows must be taken into consideration in the debates over natural light, outside air, and window proximity. Some trading off of desirable outcomes is required in order to plan and design windows that meet the needs of building users with reference to their functional comfort and the type of work they are performing.

IMPROVING LIGHTING COMFORT

In Chapter 4, we saw BIU results from a building which showed a poor air quality rating which data interpretation attributed to uncomfortable lighting. Occupants of this building spent most of their workday at computer screens where they processed invoices, dealt with customers, and managed sales. In many parts of the building, these workgroups were densely configured and equipment-heavy and worked under stressful conditions of noise, low privacy, and heat buildup from equipment. Low lighting comfort ratings were received from those areas of the building with heaviest screen use; these areas had the same lighting as all other areas of the building, namely an overhead four-bulb fixture with flat acrylic lens. This fixture provides enough light to illuminate close visual tasks on the desktop but also created significant glare to workers on screens. In response to the Building-In-Use Assessment, the building manager reduced for four-light fixtures to two bulbs, using cool-white rather than warm-white lamps to avoid creating an underlit visual environment.

This example gives an indication of the potential of the Lighting Comfort Opportunity. If such a small change can contribute so effectively to improving workers' functional comfort, then it follows that appropriate lighting technology can substantially increase workers' performance. Although the advent of CRT use on a massive scale has done much to attract attention to lighting functional comfort and lighting design, naïve tenants are still moving into office buildings built with standardized lighting fixtures in regular overhead formations that show little sensitivity to individual or task differences. Many office workers profess a dislike for conventional fluorescent lights, and yet this is by far the most common lighting installation in North American buildings. A wide range of alternative lighting is available, including various types of fluorescent, high intensity discharge, and halogen lamps, in a variety of configurations that can be adapted to a range of different task requirements. Better lighting design at an early stage of occupancy means fewer complaints from workers, better work performance, and savings on operating costs over the life of the building. Innovative lighting technology may cost more to install initially, but its value does not depreciate and energy costs may be significantly reduced.

In summary, considering the functional comfort importance of lighting comfort and the potential energy savings of an efficient system, taking advantage of the Lighting Comfort Opportunity is a logical step for most cost-conscious companies. The standardized, uniformly bright ceiling systems provided in most office buildings should no longer be acceptable to tenants or owners. Functionally comfortable lighting design maintains a bright, colorful, variable visual environment, provides adequate light for reading and writing, protects users from glare in the visual surround as well as glare on screens, and provides individual control over local and individual lighting that will optimize contrast conditions and minimize glare. Although an innovative approach to lighting requires a capital investment on a scale approaching new furniture or carpeting, it is potentially a better payback opportunity, especially if, as occupants' feedback leads us to surmise, it causes occupants to reduce their concerns and complaints about indoor air quality.

The next chapter discusses a functional comfort category that has been added to the seven key BIU dimensions because of the prevalence of concerns, comments and complaints by users and managers about these issues. The concept of "Building Convenience" comprises those characteristics of the building that affect people using the building, but are not directly related to the performance of work. However, building convenience issues have important cost implications for decision-makers, who have ultimate control over the quality of building convenience in the marketplace.

NOTES AND REFERENCES

1. L. Harris, and Associates, The Office Environment Index, (Grand Rapids, MI: Steelcase Inc. 1987); and *Stage One Total Building Performance*, (Ottawa: Public Works Canada 1985).
2. Lighting controls come in the form of individual switching in offices, as well as lights on dimmers, adjustable and movable task lights, light-or movement-sensitive cells to

regulate when lights go on and off; windows can be fitted out with a wide variety of adjustable coverings that enable people to receive the amount of daylight they need without engendering glare.

3. *The Lighting Handbook,* Illumination Engineering Society, 1993.
4. Alan Hedge, "Design Innovations In Office Environments," in W. Preiser, J. Vischer, and E. White eds., *Design Intervention: Toward a More Humane Architecture,* (New York: Van Nostrand Reinhold, 1991), p. 301.
5. Dr. Dale Tiller, personal communication.
6. The light levels specified for these buildings was 700–800 Lux—about 70 foot-candles.
7. Hedge, "Design Innovations In Office Environments," p. 301.
8. A.J. Wilkins, I. Nimmo-Smith, A.I. Slater, and L. Bedocs, "Fluorescent Lighting, Headaches and Eyestrain," *Lighting Research Technology,* 21 (1), 1989, pp. 11–18.
9. Alan Hedge "Lighting the Computerized Office: A Comparative Field Study of a Lensed-Indirect Uplighting System and a Parabolic Downlighting System" Department of Design and Environmental Analysis, Cornell University, Ithaca N. Y.: October, 1989.
10. Belinda L. Collins, Will Fisher, Gary Gillette, and Robert W. Marans, "Second-Level Post-Occupancy Evaluation Analysis" *Journal of the Illuminating Engineering Society,* Summer 1990, pp. 21–44.
11. J. Vietsch, D.W. Hine, R. Gifford, "End Users' Knowledge, Beliefs and Preferences For Lighting," *Journal of Interior Design,* 19 (2), 1993, pp. 15–26.
12. W.B. Delaney, P.C. Hughes, J.F. McNells, J.F. Sarver, and T.F. Soules, "An examination of visual clarity with high color rendering fluorescent light sources" (paper presented at I.E.S.Conference, New York, August 1977); and B.L. Collins, "Evaluation of the Role of Luminance Distributions in Occupant Response to Lighting" (Paper presented at C.I.B.S.E. National Lighting Conference, 1990): Lighting Group, National Institute of Students and Technology, Gaithersberg, MD.
13. Richard J. Wurtman, "Biological Implications of Artificial Illumination" (Paper presented at National Technical Conference of Illumination Engineering Society, Phoenix, Arizona, September 1968).

CHAPTER 8

BUILDING CONVENIENCE AND BUILDING AMENITIES

"There is an odor in the men's room. They do not fix it but they do replace the toilet-paper. What I want to know is, what do they do with all the half-used rolls of toilet-paper that they remove?"

Building occupant

WHAT IS BUILDING CONVENIENCE?

Building convenience addresses those aspects of building use that are important to people for other reasons: they help save time, they make the building attractive, or they provide that "little bit extra" that make people's lives easier. Building convenience is not the same as functional comfort because it does not directly impact the performance of work by the individual or workgroup, but building convenience issues arise repeatedly when occupants provide feedback about their buildings, because when these elements of the physical environment do not work, they create a real sense of inconvenience. Primary among building convenience issues are the following:

- car access, public transportation, and parking
- bathroom access, size, location, and hygiene
- elevator access, frequency, and maintenance
- maintenance and repair in the building

- environmental issues and energy management
- security and safety issues
- electrical power availability and accessibility.

Building convenience also addresses building amenities, such as cafeterias, outdoor areas, fitness facilities, daycare, copy centers, and a plant-filled atrium or sometimes a roof garden. If these amenities are available, they need to be accessible, well-maintained, and cheap or free to building users. If they are not, people prefer not to have them at all.

If building convenience aspects of the building are satisfactory, few people notice them. If, on the other hand, they operate poorly, users react with well-deserved outrage. Building convenience issues can often consume disproportionate amounts of management time—especially building managers—in spite of not being directly related to the performance of work, because they are related to areas such as maintenance, cleaning, security, and electrical power: the traditional core of building management responsibilities.

Although on the face of it mundane, building convenience has an important role in the O–A relationship. First, building convenience is the bottom line in terms of the quality of space that tenants should get from owners and that owners should accept from developers when they take over a building. Many of the failures in building convenience are costly, time-consuming to repair, and generate financial losses over the lifetime of the building that are not anticipated by either tenants or owners. In inconvenient buildings, businesses end up spending money fixing the basics—a *cost* to the company—which they could be spending on the functional comfort of employees—a *benefit* to the company. And second, building inconveniences, regardless of their cause, have a dramatically negative impact on the relationship between building users, who might be company employees (in owner-occupied buildings), or might be tenants in leased space, and their building managers. If this relationship is bad, communication will not take place and environmental negotiation will not be able to contribute to employees' functional comfort. For example, people will negotiate their lighting, air quality and furniture; they will not negotiate bathroom hygiene or elevator speed and convenience. It is, therefore, important to maintain the quality of building convenience.

BUILDING-IN-USE ASSESSMENT
OF BUILDING CONVENIENCE

Occupants freely volunteer feedback on building convenience regardless of the way in which information is acquired. As part of Building-In-Use Assessment, feedback on building convenience is provided by respondents in answer to a request for additional comments not covered by the rating scales on the BIU questionnaire. The predictability of the issues and concerns raised voluntarily by respondents confirms the existence of building convenience as a construct. Although each building claims different priorities, in building after building, the same topics arise.[1]

Building convenience issues affect employee morale and efficiency due to the impact of the building as a whole on the process of getting work done. Unlike functional comfort, however, building convenience affects all building occupants more or less equally, regardless of where they are located in a building and of the work they are doing there. Each building convenience topic is addressed below.

Bathrooms

Complaints about bathrooms abound—understandably, as office workers use them several times a day. Typical problems such as poor maintenance and hygiene, poor ventilation, occasional flooding, failure to replenish supplies, too small, and too far away, have an impact on employee productivity in both direct and indirect ways. In one building, customer service employees who work on telephone and computer equipment that dictates strictly observed times for coffee and lunch breaks have bathrooms with air dryers rather than paper towels, and only one dryer is available. This means that as all members of each workgroup uses the bathroom at the same time, there is inevitably a line up to dry hands, and people either return late to their desks or lose time from their break waiting for the dryer to become free. Although not a big issue to building or workgroup managers, this wait 3–4 times a day is highly frustrating to employees, whose productivity is measured in terms of the number of telephone calls handled daily and whose work time is therefore diminished on a regular basis by such a small and irritating inconvenience. In another, more serious example, employees using toilets that regularly failed to flush owing to a water pressure problem in the building often ended up by stuffing paper towels into the toilets. The ensuing plumbing problems were expensive to fix for building staff and inconvenient to employees when toilets had to be closed.

These problems adversely affect the relationship between occupants and managers. There is no reason why more than one dryer could not be installed in the bathrooms. And while occupants may not understand low water pressure problems, they do understand odors and flooding in bathrooms, and they view the lack of building convenience as poor service. This view increases the antagonism in an already adversarial relationship where managers see employees as wantonly stuffing toilets, and employees see managers as failing to maintain bathrooms. Moreover, the appearance of poor maintenance and a lack of bathroom supplies and dryers have the effect of lowering employee morale. A neglect of such detail gives employees the message that their convenience does not matter if dollar expenditures are involved. This message fuels rather than defuses user-manager antagonism.

Elevators

The range of complaints about elevators is small, but not the number. People dislike the same things, where they occur, in every building. Complants include: elevators are too slow (the wait is too long), doors close too fast, they are often dirty inside, they are often out of service, and, occasionally but forcefully, people should not be allowed to smoke in them.

The most serious problem for worker effectiveness is slow elevators, as these measurably reduce the amount of time people spend at their desks. Almost everyone agrees that time spent waiting for and inside elevators is lost time; little else can be done while waiting to get to work. Renovated buildings and new buildings which have replaced corridors (horizontal circulation) with elevators (vertical circulation) have been known to reduce staff communication by reducing opportunities for informal encounters among staff. People who meet in corridors and hallways will stop to greet and talk to each other, often solving problems that would otherwise have required more formal communications. Elevators do not provide the same opportunities for conversation, and even if people meet in elevators, these opportunities are neither private enough nor offer time enough for people to communicate significant information.

Although elevator renovations are expensive, improving speed and reducing waiting time, slowing down doors that close too fast, and maintaining them better so that they are less often out-of-service and also cleaner and more attractive, are clearly investments in occupants' comfort and convenience that should constitute part of good quality service provided by building owners and landlords to their tenants. For tenants that receive their clients in their offices, such as law firms and certain government agencies, the elevator is often the client's first introduction to the building and as such constitutes a revenue-related element of building convenience for tenant companies.

Access, Parking and Transportation

Problems in this category have to do with the physical environment of the garage (too dark, too cramped, too many speedbumps, hard to access) or parking (not enough or too expensive). Complaints can extend to the public transportation system and, therefore, the location of the office building, especially if this is not downtown and workers have to rely on irregular public transportation or bring their cars.

Parking and/or problems of public transportation are issues about which facility managers, line managers and employees themselves can do little. Because managers have little interest in collecting occupants' opinions on topics about which nothing can be done, it is therefore one of the topics on which feedback is not routinely sought. Again, as with bathrooms, longterm misfit in this area augments manager-occupant antagonism rather than improves communication. The only aspect of garage/parking as a topic on which managers can take action is safety and security of parking areas. Other access and transportation problems, such as insufficient parking spaces and inadequate public transportation, will make themselves known to those in the company responsible for site selection and may eventually lead to a move.

Safety and Security

This area is clearly one of those areas of concern that do not exist unless there is a problem. Most North American office buildings are considered safe by occupants, especially in terms of accidents or threats from outsiders. They are not considered so safe in terms of the longterm impact on occupants' health.

When security issues are raised, they are often in terms of areas such as the parking garage, outside areas at night, and building access in bad weather. Inside the office space, security comments most often made concern thefts of purses and fear of working at night. The need to feel secure is basic to human beings, and managers do not hesitate to make security issues a priority, especially as most security measures are behind-the-scenes actions that conform to managers' view of themselves as technicians whose job is to keep the building running smoothly. Security measures are an important area where managers can communicate more extensively with occupants over actions that have been taken to protect them, and where occupants can take on more responsibility for their own safety through their own behavior.

Security is basic to the point where it can hardly be defined as a need. If security is not there, work cannot take place and indeed people would stop coming into a building at all. In an office building in Boston occupied by state government offices, and therefore unguarded and open to the public, notices were posted in corridors on the main floor warning women against entering certain hallways and lobbies alone. Occupant surveys were not necessary to know that morale was low, absenteeism was high, and workers were actively campaigning to be moved out of the offices in this building.

Cleaning and Maintenance

People are critical of anything that appears to look like insufficient care of the building, perhaps because it implies an insufficient concern for the employees who work there. People dislike dust on their desks, full trash containers, litter in the entranceways, unemptied ashtrays, dirty bathrooms and elevators, and anything that goes unrepaired, such as broken doorhandles, squeaky hinges, flickering lights, and running water in bathrooms.

A recent customer satisfaction survey (not a Building-In-Use Assessment) which sampled occupants of some eleven buildings found that after indoor air quality and thermal comfort, people were most critical of poor dusting and vacuuming of their offices. Sometimes, conditions such as these are the direct result of budget reductions, where cost-cutting measures cause vacuuming of carpets, for example, to drop from three times a week to once a week. Sometimes they are a result of occupants' own behavior, as, for example, where people insist on eating at their desks, thus leaving food remains in their garbage which cause bad odors as well as attracting pests, or where people leave their desks or the floor too cluttered to be cleaned.

A typical problem occurs when buildings have to accommodate 24-hour shift work, and cleaning and janitorial services need to be available around the clock. In one building studied, where workers occupy only one floor on a 24-hour basis, cleaning and maintenance personnel are not available between 5 pm and 7 am, so that although facilities are in use, carpets are not being vacuumed, garbage is not being removed, and bathroom supplies are not being replenished. This issue caused more negative comments and criticism than any other single environmental feature on the floor.

Occupants often demand instantaneous responses to their reports of maintenance issues, such as burnt out lights, broken equipment or furniture, temperature problems, or other items that need to be fixed. If the response is slow or ineffective, occupants consider they are receiving poor quality service and an adversarial occupant-manager relationship develops. Cleaning and maintenance issues therefore benefit from an increase in communication and negotiation between occupants and managers. If senior management reduces the cleaning and dusting budget, managers need to know how important this is to occupants and how affected they are by reduced cleaning and maintenance at a functional comfort level. Because managers need occupants' feedback in order to balance cost reductions with quality increases, they need to address occupants directly and to facilitate the exchange of information between occupants and managers.

Energy management and environmental issues

While occupants are increasingly aware of energy issues, they usually know little about the energy-conserving measures that have been taken in the buildings they occupy. On the other hand, occupants often express concern about evidence of energy waste, such as lights left on in an empty building, computers and other equipment that is left running when not in use, and overly warm indoor temperatures.

There is evidence that occupants themselves could do more to conserve energy if they were better informed. Widespread wasteful behaviors, such as placing papers and documents on top of heating/ventilation units under the window blocking airflow and temperature regulation, can lead to mechanical systems imbalance that is costly in energy terms. If they are made aware of the consequences of their actions, most people will think twice before playing with thermostats or taking other thoughtless actions that waste energy.

Energy management, along with other environmental issues, is an obvious opportunity for managers and occupants to work as a team to improve building quality. Bell Canada, for example, has had not only a major success with its *Zero Waste* program, in which it has reduced the trash thrown out of its buildings by more than 50 percent in one year, but has also received systematic feedback from employees that they are both proud of and satisfied with the program.[2] Energy management and environmental issues are potentially valuable areas of occupant-manager communication, as much wasteful behavior could be controlled with more information to occupants, and workers themselves are growing increasingly aware of environmental issues.

MEASURING BUILDING CONVENIENCE

The extensive commonality of content of the comments volunteered by respondents in all kinds of office buildings across North America suggests that measurement scales can be added to the standardized BIU questionnaire survey to allow scores to be computed for building convenience. However, a normative score for building convenience is not the equivalent to the seven functional comfort scores because people only notice building convenience issues when something is amiss. As a result, there can be no positive rating for building convenience: it is either neutral or it is negative. This is not to say it is unimportant, however, because the absence of a negative rating for building convenience is a positive result.

There are, in addition, certain issues about which facilities and building managers receive feedback which do not pertain directly to the work environment or even the building. These include complaints about the telephone system, the E-mail system, or about corporate policies and procedures that affect how work is done but are not obviously building related. In many cases, these issues are raised by occupants because they do not understand or do not have adequate information on an issue. Complaints about a new telephone system in one building, for example, were indicative of people's unfamiliarity with it and lack of information on how to use it. In many cases, the most appropriate response to comments about non-building-related issues is wider and more effective information dissemination. As facilities managers become more involved in the business strategy of the organization, so their role as information managers and disseminators becomes more important.

Feedback from occupants on the seven BIU dimensions also indicates something about certain aspects of building convenience, for example, building hygiene and maintenance. If chillers have not been properly maintained and stop working during the summer when air conditioning is needed, for example, people will report poor thermal comfort and air quality conditions. If occupants' lights are burned out or flickering, this will be reflected in their lighting comfort ratings.

Logically, however, building convenience should not have to be measured: it should exist. This dimension is in contrast to functional comfort, which is by definition negotiable as task requirements change, people's needs change, and the interaction between workers and their environment is fine-tuned. Building convenience should not be a negotiable commodity: it is a basic building block on which environmental quality is constructed. If future building design and construction are to be acceptable in the market place, more effort needs to be invested in ensuring building convenience.

BUILDING AMENITIES AND THE FUTURE OF BUILDING CONVENIENCE

The amenities offered in a building traditionally affect rents that landlords can charge as well as investments companies make in the wellbeing of their employees. For example, cafeterias, fitness centers, shopping opportunities, adequate parking, and daycare centers are often found inside modern office buildings and can have a significant effect on occupants' ratings. Other buildings, by virtue of a downtown location, for example, do not include amenities for occupants because so much is available nearby. People's ratings of a building's amenities are therefore not always as related to the number of amenities provided by a building as they are to the amenities of the building's location.

In one BIU Assessment, all seven of the BIU dimensions received strongly positive scores, indicating that people's functional comfort was adequate in the building. However, their comments suggested that the employees did not much like their building because of the lack of amenities outside the office space. Their demands were significant: a Post Office, a shop, more parking, an outside area, and a fitness center. Facilities managers can do little to provide amenities at this scale if they are not already available in the building or if senior management has not already decided that investing in a fitness room because a health center reduces health premiums, raises morale, reduces accidents and illness, for example, is worth the expense.

As employers start to reexamine the workspace, however, and to consider cost-effective alternatives to the downtown office tower, environmental elements previously considered as amenities and therefore as additions to the basic workspace are coming to be considered as necessary elements of the work environment. A software company near Montreal has provided employees with a swimming pool; in Japan, Kajima Corporation seeds its ventilation system with attractive fragrances; Digital Equipment Corporation in Finland has provided its sales teams with a porch swing; and the NMB Headquarters in Amsterdam has scattered coffee shops, restaurants, waterfalls, and ornamental pools throughout the building. All three are examples of highly successful enterprises that focus on understanding and defining how their most creative people work, and have made the officespace into the environment which helps and supports the work that they do in the way they do it best.

In the first example, the employees are highly trained electronics engineers who work day and night when trying to solve a problem. The pool not only helps them integrate recreation with work (or vice versa) but also attracts their families so that spouses and children are entertained while the engineer debugs his/her program. In the second case, some fragrances are thought to increase alertness, whereas others help people relax. These are wafted to office floors and to the employees' lounge, respectively. As reported in Chapter 1, the sales staff in DEC's Finland offices are rarely in the office as they are mostly on the road, and in lieu of individual desks, they opted for an especially attractive common space in which they could talk, plan, exchange information, and receive clients. The porch furniture is an important element of this process, and the porch swing is not considered an amenity so much as a necessary piece of office furniture. And in Amsterdam, the Bank's employees perform their clerical and administrative tasks in office space designed to optimize functional comfort. For coffee and meal breaks, meetings, group sessions, and other activities that take place away from the desk, the environment provided is deliberately un-office-like, with massive indoor plants, pastel colored finishes and colorful artwork, in addition to the flowing water. Thus employees' breaks from the individual workspace take place in an environment that is designed to soothe, comfort and please while creative work goes on.[3]

These examples are among many that demonstrate that the distinction between building amenities which have been added by developers to increase the profitability and market value of speculatively built real estate, and the functionally comfortable workspace—which is no longer limited to individual desks in offices—is dissolving. By the same token, the distinction between the comfort of basic building habitability, as connoted by the concept of building convenience, and the functional comfort of employees, is one that should be strengthened. Occupants of office buildings should not have to be inconvenienced by basic needs such as bathrooms, elevators, and security: tomorrow's cost-conscious and streamlined organizations will pay their rent in return for space in a building that is guaranteed to perform for them. They will not spend money solving problems caused by insufficient water pressure, slow elevators, unsafe parking areas, and excessive energy consumption. At the same time, tomorrow's organizations will no longer evaluate building amenities as extras or additions to the basic office space. They will respond to their employees' demands for variety, comfort, choice, and flexibility by integrating environmental elements formerly considered amenities with traditional workspace requirements.

In summary, existing trends imply that the workspace of the future will:

- provide a range and variety of group spaces more imaginative than conference rooms and lounges to facilitate teamwork, project work and meetings of all sizes and types,
- contain opportunities that in the past have been considered recreational and separated from work in recognition that the most productive and creative individuals do not turn themselves on at 9 and off at 5,
- respond to the idiosyncrasies and task variability of different workgroups by providing nontraditional elements in the work environment.

In the future, corporate managers seeking a more aggressive return on their investment in accommodation will demand better quality buildings from suppliers. Most cities have overbuilt their office space supply, and as a result building owners are competing for buyers and tenants. Those that will realize the most profit will provide a guaranteed level of building convenience resulting not only in lower operating costs over time for buyers and tenants, but also in less staff time invested in building maintenance and repair. Today's top-of-the-line office buildings are expensively finished and aesthetic to look at; they may or may not work properly. Tomorrow's top-of-the-line office buildings will be environmentally responsible and energy efficient, and they *will* work properly. As a result, the focus of corporate expenditure of time and money on accommodation will be on employees' functional comfort and the effectiveness of the workspace rather than on basic building maintenance and repairs.

In response to these trends, building convenience takes on a new meaning—that of the basic functionality of whatever spaces are being occupied. The attitude hitherto accepted in business—of selecting buildings on the basis of cost, location and image, and of paying facilities staff to make sure things work—needs to be replaced by an informed demand from consumers for overall better quality building stock. Companies no longer want to fight their buildings to get good accommodation for their employees: bathrooms must work, elevators must be prompt, garages must be safe, and the building should be environmentally responsible. This taken care of, knowledgeable managers are freed up to negotiate functionally comfortable workspace for employees.

Additional functional comfort dimensions may be developed as work environments other than conventional offices are examined for feedback from their occupants, such as school classrooms, hospital wards, and industrial environments. In these more complex buildings, additional groups of occupants can be defined beyond the employees of the organization, and therefore feedback from a variety of users collected. Those who visit the space, those who operate and maintain it, and those in whose interests the space exists (such as hospital patients) are as much users as those who work in the space. In such buildings, therefore, the viewpoints of different groups of users may not be the same, and their requirements from the building must be reconciled. This complexity added to the environmental negotiation process reinforces the importance of people in the organization other than facilities managers in taking on responsibility for their physical environment. Once these responsibilities are more fully shared in organizations, a broader range of consumers of space in buildings will create an effective demand for better quality buildings and better quality building technology.

In this chapter and the three that preceded it, much attention is given to understanding how people behave in buildings and how this knowledge can be linked to the real estate mission and ultimately to the business strategy of the organization. But what about the process? How do companies apply the work environment feedback they acquire from their employees to decisions pertaining both to the management of assets and property, and to the goals and mission of the organization? In the three case studies presented in the next chapter, three different ways of answering these questions are offered. Each of the three organizations differs in its style of work, its corporate mission and its mode of operations; they do not differ, however, in their desire to be profitable, to impress their shareholders, and to show evidence of success to them and to the world. And with varying degrees of success, each has used Building-In-Use Assessment to improve their O–A relationship to do so.

NOTES AND REFERENCES

1. In recent versions of the BIU questionnaire, additional rating scales are used to predict scores on a building convenience factor, but no norms have yet been calculated.
2. S. Quesnel, "Zero Waste," (Paper delivered at IFMA Utilities Conference, Montreal, May, 1993).
3. Vischer and Mees, "Organic Design in the Netherlands," p. 285.

CHAPTER 9

OCCUPANTS' FEEDBACK AS A DECISION-MAKING TOOL: THREE CASE STUDIES[1]

"All human beings—not only professional practitioners—need to become competent in taking action and simultaneously reflecting on this action to learn from it."

C. Argyris and D. Schön, *Theory in Practice*

ORGANIZATIONAL LEARNING THROUGH THE ACQUISITION OF FEEDBACK

Part of what is important about feedback from users about their accommodation is the opportunity it provides for an organization to learn, in order to change. In learning about how its members use space, an organization can make better decisions about—and can derive more benefit from—the space it occupies. In all organizations, there are forces that favor change and forces that want to see things stay the same; organizations that do not take advantage of opportunities for learning eventually wither and die.

In their analysis of "organizational learning," two organizational theorists, Argyris and Schön, distinguish between "single-loop learning" and "double-loop learning."[2] Pointing out that an organization is more than the sum of its members, and that individual learning does not substitute for organizational learning, they define single-loop learning as occurring when

> members of the organization respond to changes in the internal and external environments of the organization by detecting errors which they then

correct so as to maintain the central features of the organizational theory-in-use.[3]

Although single loop learning solves problems and improves organizational functioning, it does not fundamentally change the organization's values, attitudes, and mode of operation (its theory-in-use). These changes occur in "double-loop learning:"

> those sorts of organizational inquiry which resolve incompatible organizational norms by setting new priorities and weightings of norms, or by restructuring the norms themselves together with associated strategies and assumptions. [4]

Both types of learning involve inquiry and solve conflicts, but in single-loop learning, these actions are carried out to enable the organization to continue functioning in the same way, without challenging fundamental norms and assumptions. Double-loop learning occurs when the organization learns to change profoundly its way of doing things—its theory-in-use —in order to solve the conflicts or problems that have arisen.

In the case studies discussed in this chapter, organizational inquiry occurred because the company wanted to develop a better understanding of users' functional comfort, and of how well its accommodation performs. In these cases, there was no identifiable problem needing to be solved; there was a desire to improve, to do better, and to learn. Each organization described here first sought to acquire a decision-making tool to facilitate single-loop learning, but then found out how difficult it was to learn from the information it acquired because this would have meant radical organizational change, more like double-loop learning. In the first case study, the organization's theory-in-use allowed a separate and distinct facilities department to operate by following orders and instructions issued by business units and by providing services on an as-needed basis: the "Engineering approach" in terms of the model described in Chapter 2. The feedback from users was used to improve service, and, in fact, some individual learning accrued to facilities staff members from acquiring occupant feedback about the work environment. Some single-loop learning also occurred as the facilities organization used inquiry to respond to external challenges. There was, however, no double-loop learning; the organization could not change its way of doing things, and although it solved air quality, lighting, and furniture problems, it resisted all the pressures for internal change created by the occupant feedback.

In the second case, inquiry into feedback on space use led more readily to single-loop learning in that the organization's theory-in-use already incorporated the need and desire for change. This organization was actively seeking ways to change and improve its behavior in response to the increasing competition in its business environment. It used occupants' feedback to solve some of the problems its employees had in their buildings, and therefore to improve the O–A relationship. Although aware that double-loop learning moves an organization towards changing its habits and norms, this particular organization could not ultimately manage to avoid the conflict generated when feedback from building users was applied to the established space decision-making process.

In the third case, accommodation services are the business of the organization, thus its inquiry into the accommodation adequacy of its clients was predicated on the desire to learn from their feedback, to increase competitive advantage, and to improve market share. It was largely able to meet these objectives without challenging organizational norms and assumptions. It, therefore, cannot be said that double-loop learning occurred in this case. The company has not yet fully integrated the innovation of user feedback into its mode of operation, so the need for conflict resolution may increase in the future, at which time more profound organizational change may take place.

These three organizations, having systematically acquired occupant feedback from their buildings, have used it for learning in a variety of ways to change work environment conditions, solve building problems, and improve communication. In this chapter, the three examples of how the information has been used are presented along with a discussion of their effectiveness, the resources required in each case, and the overall impact on the organization. All the case studies demonstrate the impact of information on the facilities organization, on the organization as a whole, and on decision-making processes.

In organizations that have no way to apply and follow up on occupancy feedback, it is not surprising that occupant surveys are often disparaged as a cosmetic and ineffectual tool. The cases presented in this chapter show that although the quality of feedback can be good and the information valid and reliable, the major organizational challenge is not so much to acquire feedback from occupants as to devise appropriate applications for feedback about the environment, channeling the results to the right people, and facilitating the decision-making processes the information is supposed to inform. Information alone does not solve problems, and can even cause them. As Senge points out, using information to repair symptoms of problems is not as effective as taking actions to change the structures within which problems occur so as to "lead to significant enduring improvement."[5] The case studies illustrate how, when an organization decides feedback is needed, it has to take extensive steps to process, digest, and apply that information if the process is to meet its needs. Acquiring the information without a specific purpose or context to which it can be applied, although not entirely fruitless, often reduces significantly the value of the undertaking. Information is only a valuable tool in an organization if it can be applied to test a theory, solve a problem, or otherwise be applied to action.[6]

In the first example, the facilities group of a large, quasi-governmental international organization (International Co.) is a positive and experienced user of occupancy feedback from its buildings. Its managers welcomed employee input, but they had trouble responding systematically to the data and tended to let it fall by the wayside. Employees were rather jaded about being asked for their opinions, as a result. In the second case, the facilities staff of a large telecommunications organization (Telecom Co.) wanted to use occupancy feedback to create open dialogue with building occupants. They enthusiastically collected information but found they had almost no mechanisms in place for opening up such a dialogue. In both cases, more than simply collecting, analyzing, and acquiring information was necessary to meet organizational learning objectives. In the third case study, the organization was itself a property management firm (Comre Co.), and occupancy feedback from tenants was expected to assist staff in making better decisions—but were staff ready to use this tool?

INTERNATIONAL HEADQUARTERS OFFICES:
Conflicts Over Information Utilization

International Co. is a large international organization with headquarters in Washington DC which occupies all or part of some eight large office buildings downtown. With a history of concern for users' needs and employee involvement in planning and programming, the facilities organization has carried out a series of inquiries into functional comfort as well as into customer satisfaction in its buildings. International Co. is a large complex organization whose mandate and mission require it to adapt continually to changing political conditions worldwide. The organization's major challenge regarding occupancy feedback is to integrate information on building comfort and functionality into an ongoing process of complex and multilevel decisionmaking. Previous efforts by facilities staff to be responsive to clients' needs have often been rendered ineffectual by the overriding priorities of the organization's business mission. Decisions to move departments are changed at the last minute; space allocations are challenged by departments who feel they are being cheated of space; space and size standards are often violated by executives for themselves or for their personnel; and information about departments' future plans and changes is not always provided to facilities planners who are responsible for space needs forecasting.

The facilities organization of International Co. comprises four interactive and overlapping groups of varying size, as diagrammed in Figure 9.1. The largest staff and budget allocation belongs to the group responsible for building maintenance, repair, and operation, as well as for maintaining mechanical and electrical systems in owned and leased buildings, and for janitorial and cleaning services. The "Service Center," which fields complaint calls and issues service orders for repairs and changes, is located within this 25-person group; this group also carries out the construction of new office-space when groups move within company buildings. The space planning and design group coordinates and designs space for departments making such moves. Some members of this group are responsible for the overall allocation of space to workgroups. Others coordinate input from users with space planning decisions, and interface with communications systems, and engineering personnel to produce space plans and moving schedules; they are also

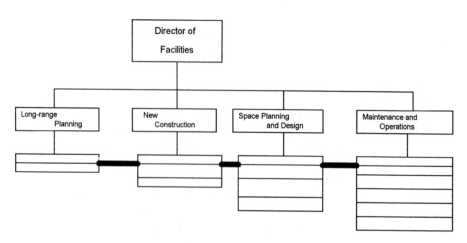

Figure 9.1. Schematic organization chart of the facilities department of International Co.

responsible for keeping building drawings up to date. The other two groups are responsible for leasing space and for the design and construction of new buildings, respectively.

Occupancy surveys are a regular feature of life in International Co. Job attitude surveys, food services surveys, and space-use surveys, among others, are frequent. In the facilities organization, feedback from occupants is considered both necessary and important to ensure a high quality work environment for employees as well as to provide good quality services. Building-In-Use Assessments were carried out over a period of two years in four owned and four leased buildings in downtown Washington, ranging in size from 8–12 stories and accommodating anywhere from 200 to 1,000 occupants. The buildings are of varying age, but the older ones have been upgraded so that all buildings maintain a high standard of interior comfort, modern furniture, computer equipment and communications technology. The results of the Building-In-Use Assessments have been consolidated into International's own BIU database, from which normative scores for International's buildings have been computed. These scores serve as the benchmark to which each freshly surveyed building is compared.

International Co.'s design staff have developed space standards which govern office size and furniture configuration and protect access to daylight for the different ranks and jobs in the organization. Facilities staff were eager to have feedback from occupants regarding these and related space-use decisions. They wanted to use occupants' feedback to strengthen the link between accommodation services and International Co.'s business mission: if people were appropriately and effectively accommodated, they would provide more effective services to their own clients. Over time, members of the facilities staff learned to execute and to analyze data from occupant surveys themselves. They learned to apply the results to a series of actions and changes within the organization: before and after measurements of functional comfort, diagnosis of the precise location of indoor air quality and thermal comfort problems, and applications to the design of new space.

At first glance, International Co. appeared to have implemented a fluent and successful occupancy feedback system, in which information was collected, analyzed, and used unabashedly to solve problems by the facilities organization. But closer inspection revealed that although mountains of data were being gathered, information was not flowing anywhere. In fact, several major blockages existed—some procedural, some political, and some concealed—to the flow of information and its application to decision making. So although feedback was being gathered from building occupants, the organizational changes that would have been necessary to make it effective as a decision-making tool were not taking place. Individual planners and designers used the feedback to solve problems and change their way of doing things, but the norms and attitudes of the organization—inside and outside the facilities unit—were not ready to change.

The Need For Communication

In deciding to consolidate BIU survey data into its own BIU database, International Co.'s facilities staff first had to define how to use it as a tool. There was some question as to whether it was more useful to compare each building's BIU results to other North American buildings, thus comparing International Co. with industry, or whether to compare each building to its own (high) standards. Rather than defining how the database was to be used, however, a political issue surfaced. Some thought that by comparing International Co.'s buildings to other North American buildings, they could show the employees what a good environment they had and how successful they (facilities staff) were in providing a comfortable work environment. Others felt that it would be more useful to know how the buildings compared with each other inside the organization; they could, then, rapidly determine which of the buildings they were responsible for were successful and which needed more attention, and they could use the database and the norms as a decision-making tool. Thus, although significant amounts of feedback from occupants was already available, a conflict existed regarding how to make occupancy feedback work as a tool, with the first group favoring a more reactive approach, in which occupancy feedback would be used to confirm an existing pattern of facilities activity and decision-making, and the second group seeking to compare themselves with peers to provide guidance on how to improve and change.

Controversy also erupted over how to process, where to direct, and who should follow up on the action recommendations resulting from the surveys. No one group wanted to take responsibility for this process, as the data impacted all the facilities groups and to some extent, other work groups in the organization. No one could agree on how (or if) action recommendations should be documented, what mechanism could or should be implemented to ensure they were carried out, and how to inform occupants about the follow-up to their survey data. Had steps been taken to resolve these conflicts, single-loop learning could have occurred through facilities staff applying occupants' feedback to problem-solving. This single-loop learning might eventually have led to double-loop learning by causing the facilities group to move to change the organizational norms that kept it in a reactive and subordinate role in relation to the larger organization.

In fact, facilities staff opted to solve the first conflict by comparing individual building ratings with both in-house *and* industry norms. With only seven scores, it appears to be a simple matter to make two comparisons, one with the industry benchmark and one with the organization's own benchmark. However, this apparently simple solution further confounded the second dilemma: to whom should the information be funneled, and how should they act on it? By making two comparisons instead of one, the amount of information to be ingested was exactly doubled, and a second difficult decision had to be taken on which of the two comparisons should be used as the basis for follow-up action.

Once the BIU database was in place, these conflicts effectively paralyzed facilities decision making: who should access the database, how to use it, and how to structure and maintain it became questions that no one could answer. The different arms of the facilities organization had different needs. For example, the team responsible for mechanical system performance wanted to have low scores pinpointed on floor plans so that they could take CO_2 monitors and temperature gauges into the exact spots in the building and measure actual temperature, humidity, and carbon dioxide levels. The design team wanted cumulative feedback correlated to location and floor layout to help them solve tricky design problems such as narrow column spacing, deep floors and windowless walls. The planners wanted scores broken out by individual work groups, wherever they were located in a building, so that they could know at a glance the functional comfort ratings of groups that were planning moves or changes. The computer-aided design team wanted the data linked to CAD; the team working on the new office building that was under construction wanted more than feedback from existing buildings, they wanted the implications spelled out for future space design. All these applications were, of course, physically possible, but none was strategically selected.

How could the follow-up action stage of occupancy feedback be managed to be less controversial? It couldn't. One facilities group wanted data interpretation with no follow-up actions; another did not even want the interpretation, stating that the data alone sufficed to interpret results and implement follow-up. Several building managers quietly used the results immediately, making building changes and corrections in response to the findings without seeking formal approval. Others wanted to be told what follow-up actions were necessary and expected before doing anything at all. But just as there was no effective process in place to guide facilities' staff's decisions about where to direct the results, so there was no effective mechanism in place to apply the results to decision-making and follow-up action.

This impasse resulted in chronic underutilization of the feedback information that had been collected. The failure to make use of their occupancy feedback had ramifications throughout the organization. First, occupants who had provided information about their work environment felt it was not heard, causing them to lose faith in the process and in the manager-occupant relationship. This loss of faith reinforced business managers' tendency to override facilities decisions and to take actions in their own spaces without consultation with facilities staff. Second, the failure showed up the intrinsic weakness of intergroup communication within the facilities organization. The ability to decide *together* how to share and use occupants' feedback is a prerequisite for effective implementation of any feedback system. None of the facilities groups' demands or expectations was incompatible, yet lack of communication caused the differences in their demands and expectations to be raised to the level of a conflict for which there was no resolution mechanism.

Organizational Impact of Building-In-Use Assessment

The BIU inquiry did have single-loop learning effects at International Co., in spite of facilities groups' problems disseminating, using and applying the occupancy feedback. One such development was the increased frequency with which small adjustments and changes were made to facilities standards and procedures as a result of feedback from occupants. For example, several buildings' BIU ratings indicated problems of acoustic privacy in enclosed offices. The construction personnel easily upgraded the standards of construction of enclosed office party walls to a higher *sound transmission coefficient* (STC). In one building, occupants seated near an elevator shaft reported odors. Instead of discounting the BIU results, the mechanical team investigated the complaint and found a leak into the elevator shaft from the kitchen exhaust in the cafeteria. They repaired it, and the odors disappeared. This improvement would never have been known but for the fact that the floor reporting this complaint was surveyed twice at a one-year interval, and the data showed that these complaints had disappeared. So, although facilities staff were in fact responding to occupants' feedback by changing their practices and procedures, these changes were "backstage" and not publicly connected to the occupant feedback process.

Systematic feedback from building occupants also changed thinking in the facilities department by demonstrating that important facilities decisions were being made without adequate technical or performance information about individual buildings. It seemed that design staff were making building changes without being fully informed by mechanical and electrical teams about the building's systems. When this lack of coordination was recognized, a building evaluation project was implemented in which a team reviewed the original plans and specifications for each building and compared these to current performance requirements in terms of actual use. Their recommendations were to be integrated with the results of the Building-In-Use Assessment to frame an action plan for each building to solve problems and improve performance with an annually allocated budget. However, the assessments were completed far more quickly than the technical evaluations.

The feedback also helped the facilities group respond to criticisms from an advocacy group representing employees' interests. This group had long claimed highhanded ignorance of people's accommodation needs by the facilities team. The team was able to point to their extensive BIU database as evidence of their interest in and investment in employees' feedback. The facilities director used his negotiations with the advocacy group to point out that there was a clear need for International Co.'s employees to become more informed about the buildings they occupy in order for them to be able to protect the quality of their own work environment. For example, business managers would inadvertently sabotage workspace quality by making changes that violated space and design standards, such as hiring consultants and giving them office space in a meeting-room, thereby losing meeting-space for staff. Facilities staff wanted employees to use the BIU results to communicate to their own managers the impact of their decisions about workspace, thereby side-stepping the employee advocacy group and the traditional adversarial relationship between building users and management. This initiative would have shifted the organization towards a more collaborative planning relationship between business and space managers, had the various facilities groups been united enough to follow it through.

In conclusion, International Co. made an ambitious start in a new direction for their O–A relationship, and there were positive outcomes for the organization. However, the facilities team alone was not a major enough player to push through the kind of changes needed, and the weight of the existing organizational culture slowed down their progress. Before embarking on another, or an expanded, program likely to bring change, the facilities decision-makers realized they need the commitment of senior executives from business operations to be successful.

NATIONAL TELECOMMUNICATIONS COMPANY: Diagnostic Information as a Tool for Strategic Planning

In the second case, by comparison, there was a significantly more pronounced corporate effort to solve conflicts created by the impact of feedback information on the organization. This large telecommunications company (Telecom Co.) has significant real estate holdings in the Northeast: 50,000 people in approximately 200 buildings, of which about 4,000 are in the company's headquarters tower. The company also owns and manages buildings for telephone equipment and for its repair and maintenance staff over a sizable geographic area. Its administrative buildings accommodate employees that provide key services to customers and support services to these employees.

Figure 9.2 shows the three groups in the real estate division responsible for the company's accommodation and property management. The design and construction staff are responsible for new space design and construction as well as for engineering and technical services. The space planning and leasing group manages space planning decisions, negotiates leases, carries out space planning for workgroups, and is responsible for leasing out or selling excess space. The facilities management and operations staff operate and maintain all owned buildings and provide services to some of the leased space. As the telecommunications business becomes more competitive, the company is taking steps to streamline its operations, reduce staff and increase productivity, and keep its market share. In real estate as a whole, the company is moving towards a revenue-dependent facilities operation, in which client groups within the organization will become financially responsible for services received, whether these are the planning and design of new space, the purchase of furniture, moving to new space, or maintenance and management of space currently occupied. Telecom Co.'s Real Estate division is trying to shift from the "Market cost and usage standards" strategy to the "Market design approach"—from Stage 3 to Stage 4 in the MIT team's model of FM financing. The amounts budgeted by each division for these services can therefore, in theory, also be spent on similar services from vendors outside the organization, if these prove to be more cost-effective. The facilities organization is adjusting to these new pressures by developing an innovative and cost-

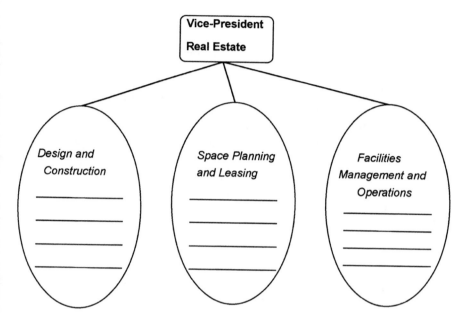

Figure 9.2. Schematic organization chart, showing facilities department of Telecom Co.

conscious approach to their operation. The facilities groups themselves have been reorganized, several services have been out-sourced, and experiments with telecommuting, waste reduction, and responsible energy management are encouraged.

The facilities team is actively seeking a more integrated role in business strategy. In recent years, they have tried to increase their profitability by reducing staff and costs, offering more streamlined services, reducing employees' space needs, increasing customer loyalty among company personnel, and making space-related decisions in innovative ways. They have initiated a total quality management program and consider feedback from their customers an important element in reaching these goals successfully. Occupants' feedback is used not only to improve services to client departments, but also to help implement innovative accommodation strategies.

Building-In-Use Assessments were initiated as a first step towards improving communication between managers and occupants and have been carried out in six of Telecom Co.'s buildings, four of which are described in this section. First, two suburban office buildings were selected because they were newly built and occupied. Both are leased, and are considered to be among the best quality office space currently provided by the organization. Each building accommodates approximately 300 employees who are engaged in offering direct services to the company's customers, that is residential and commercial telephone subscribers. Occupancy feedback was seen not so much as a tool for decision-making as to provide facilities staff with information on the appropriateness and effectiveness of their decisions.

Second, feedback was sought from two owner-occupied buildings. One is a large, older and less well-appointed office building with a history of operational problems, poor communication between building users and building management staff, and uncomfortable working conditions for employees. Telecom Co. used Building-In-Use Assessment to quantify and systematize the problems users have with the building, so that instead of a large morass of undifferentiated negativity, which is what the managers had been working with to that point, functional comfort problems could be identified separately and decisions taken to solve them one at a time. The other building is newly occupied, with a new furniture system installed on some floors. Feedback was sought on the differences in functional comfort between occupants with, and those without, the new furniture to determine whether or not to provide the new furniture system to all employees.

At a later stage, a customer satisfaction survey, and not the BIU questionnaire, was administered to a 10 percent sample of the population accommodated in all Telecom Co.'s buildings. This third survey effort provided feedback on a wide range of building-related services in addition to functional comfort. Like International Co., Telecom Co. faced the challenge of managing, targeting, and applying the feedback information to solve problems. Unlike International Co., however, Telecom Co. succeeded in evolving a mechanism for processing occupancy feedback information and directing it at those who could best use it. During the first wave of occupant surveys, facilities managers began to recognize that a mechanism was required to manage and direct results. By the second wave, facilities staff, knowing what to expect, were more receptive to the results and developed very explicit action recommendations for follow-up. By the time the customer satisfaction survey was carried out, a comprehensive system had been developed that brought the feedback to all the different levels and divisions of the facilities management organization, showing a willingness to challenge existing norms and undertaking to resolve conflicts in ways that encouraged organizational learning.

Structure of the Follow-Up Process

At each stage in the acquisition of occupant feedback, members of all three real estate groups were deeply involved in data interpretation. The team for the first two buildings surveyed used the results to diagnose specific sources of reduced functional comfort, carried out a detailed building walk-through, and eventually negotiated changes with the landlord. In the next two buildings surveyed, FM teams and a selection of space planners, engineers, senior facilities managers, and concerned staff, guided by the BIU results, engaged in a detailed building walk-through. As well as showing facilities staff how to use occupancy feedback as a diagnostic tool, the walk-throughs demonstrated to occupants that the key facilities decision-makers were paying attention to their physical space as a result of the occupant survey. These data interpretation activities were followed in each case by a series of brainstorming workshops with employee representatives to decide what follow-up actions to take and who should be responsible. The results of the workshops took the form of an action list of remedial steps documented and distributed to all facilities staff.

Telecom Co.'s brainstorming workshops allowed each facilities group to factor in its own operating constraints, to provide resources and information to each other, and to set priorities on action recommendations. For example, building managers had felt there was some risk associated with assessing the older building because of its history of building problems and user complaints. They were concerned that distributing questionnaires to occupants would unleash a torrent of complaints, that in turn would generate a backlash of suspicion along the lines of "If they are asking us about the building, there must be something wrong with it." In fact, the 6-story building accommodates 1800 people on large open floors served by two mechanical systems, and facilities staff learned through a close analysis of the scores where occupants' discomfort was most pronounced, how serious the air quality problem was, and which workgroups on which floors required most attention. The conviction that this building represented one large undifferentiated problem dissolved as a diagnosis emerged of specific problem areas, a range of types and degrees of problems, and even evidence that some areas had no problems at all.

The most negative ratings came from groups that worked all day long on computer terminals. Seated in areas densely packed with people and equipment, they reported extensive thermal comfort and lighting problems. By opening up these areas, providing ergonomic chairs, and increasing ventilation and thermostat control soon after receiving the survey results, managers dramatically reduced occupants' building complaints at minimal cost. Another outcome of the brainstorming sessions was a decision to approach Telecom Co.'s own systems' engineers to point out the significant operating and human costs of using CRT screens. Facilities staff worked with product engineers to explore alternative display technology, such as flat screens with more natural color displays that are less demanding on the human eye, as well as on building conditions (less need for special lighting, less heat generated by equipment, and less space taken up by computers).

Managers decided to disseminate survey results to occupants along with key points from the follow-up action plan. In this way, building occupants knew there was some response to their input. The written communication included references to the facilities team's budgetary constraints, pointing out that improvements could only be made in the context of senior management's lower operating cost targets for each building. Building managers used this as an opportunity to "get people on their side" by sharing with building users the problem of delivering quality services at no additional cost.

Senior management at Telecom Co. decided that the discrepancy between leased and owned office quality as evidenced in the feedback from occupants meant that to support productivity, more resources needed to be invested to upgrade their own property—the large unprofitable headquarters' building. While on the face of it *not* upgrading the space was a cost-effective decision, the advent of revenue-dependency meant that eventually the business units would trade off between low-cost, low-quality space in owned buildings, and higher cost, better quality space in leased buildings. With little additional investment, the company's own building could be rendered good quality at a lower cost than that which was available elsewhere, and the business units would spend their space dollars within, instead of outside, Telecom Co.

Facilities managers made certain that senior management were aware of the hidden costs of continuing to operate low quality space: for example, they pointed out that maintenance and repair budgets are committed no matter how good or bad the quality of the interior space, yet are less effective in improving a run-down building than a well-maintained one. Also, the cost of equipment breakdowns is higher when the equipment has to be repaired than if the equipment is being maintained which reduces the possibility of breakdowns. And constant occupant complaints and service calls to adjust thermostats, unstick elevators, patch worn carpets, and unplug toilets are costly in terms of staff time. Moreover, although there is no proof that more people cannot be squeezed into a poor quality space than a high quality one, all indications are that more people can be efficiently and comfortably accommodated over the life of the building in a well-planned good quality environment than in a poorly-maintained and under-resourced workspace. Telecom Co's business strategists eventually decided to invest in their own property and to reduce the numbers of employees accommodated in leased space. In addition, the functional comfort data from leased space provided strategists with a demonstrable functional comfort standard to meet in their own renovated space.

Telecom Co. was less ambitious than International Co. in its applications of occupant feedback. The information came into the facilities organization in reasonable chunks rather than a large unmanageable flow, and the relationship between the business management of the organization and its real estate staff is closer and more aligned in Telecom Co.—a successful monopolistic enterprise in a changing and newly competitive market—than in International Co.—a large and geographically spread out quasi-governmental service organization. Indecision about how best to use to use the feedback did not paralyze Telecom's staff as it did International's staff, but Telecom did not try to do as much with their occupancy feedback tool as did International. On the other hand, Telecom's incremental approach generated more organizational learning than did International's ambitious approach. Telecom Co. used feedback for single-loop learning by solving conflicts within its buildings. New ways to involve occupants, managers, and facilities staff in negotiating solutions to building problems resulted in cost-effective interventions that improved occupants' functional comfort and helped them work more effectively. In addition, the organization advanced towards double-loop learning by initiating an accommodation strategy planning process that changed the organization's way of making decisions. By demonstrating that money currently being spent on leases could be invested in upgrading owned buildings, and that people accommodated in good quality leased space would compete to be accommodated in just as attractive owned space, facilities staff overcame corporate resistance to investing in its own office buildings. By allying their advice with an innovative effort to move personnel out to satellite offices and home-based work, the facilities organization became strategic experts in using accommodation information to advance business strategy. Since they had to adopt new norms and attitudes to facilitate the changeover to revenue-dependency, they had an opportunity to become leaders in organizational and cultural change as the company streamlined costs and downsized. They positioned themselves to take the market design approach (Stage 4 in the FM financing model) to creative financing of facilities services. Occupants' feedback was the tool they used to keep business managers informed about the impact of accommodation decisions.

The third case study illustrates quite a different aspect of occupancy feedback—can feedback from tenants serve as a planning tool and resource for landlords, or is there a conflict of interest between property management firms' needs to fill their buildings and tenants' needs to spend dollars efficiently on space?

COMMERCIAL REAL ESTATE COMPANY:
Using Feedback to Increase Competitive Advantage

For large-scale property managers and landlords who traditionally distance themselves from their tenants' business planning decisions, the opportunity to help tenants increase their functional comfort is a way to enhance services to tenants and to replace adversarial and costly lease negotiations with a mutually supportive exchange of information. An example of this approach was used by a company, Comre Co. located in a large city in the northeastern United States which is a large investor in commercial real estate and operates a growing property management portfolio of some seven million square feet. Comre Co. prides itself on providing good quality space and a high level of service to tenants. At the time of the occupancy feedback initiative, the company was operating in a recessionary market where office space had been overbuilt and competition for new tenants was cutthroat.

Part of Comre Co.'s commitment to good service is to carry out regular client satisfaction surveys that provide feedback to facilities staff on the services they offer tenants. But the surveys have no diagnostic value in terms of measuring the functional comfort of building occupants. Comre Co. determined that in this tight market it had to offer value-added services to tenants, principal among which was assistance in planning and designing space based on feedback from tenants' employees. The company also felt that both they and their tenants would negotiate more effectively if feedback from occupants about their functional comfort were available to both sides.

Comre Co. had increased the efficiency of its property management operations by out-sourcing the technical aspects of building management while maintaining in-house lease negotiation and lease management services, and design and space planning teams. Each building was operated with a small on-site staff, chief among whom was a building administrator who managed the customer satisfaction survey and maintained a strong personal relationship with all the tenants. To initiate its occupancy feedback system, Comre Co. first executed a pilot BIU Assessment of a single building. The administrator of this building asked tenants if they would agree to take part in a diagnostic survey, the purpose of which was not just to measure customer satisfaction, but also to show how tenants' workspace could be improved. Once the survey was complete, the building administrator invited tenant representatives to discuss the findings with her. These discussions enabled the administrator to indicate not only what information Comre Co. would use to make building improvements, but also to inform tenants of what they themselves might do to improve the functional comfort of their own staff. Although one of the three participating tenants in the building expressed no interest in improving employees' functional comfort, the other two took over the results on spatial comfort, privacy, lighting comfort, and noise control. In one case, the tenant asked advice from an interior designer to improve acoustic privacy, and in another, the tenant applied the advice of Comre Co.'s own space planners to provide better spacing between employees and make a more functional space layout in their offices. All three tenants were pleasantly surprised to discover the high degree of functional comfort reported subsequently by their employees. This type of feedback on their officespace installation had not been available to them any other way.

The administrator used the BIU results to identify for herself areas of poor air quality and thermal comfort in the building, and noted the prevalence of low lighting comfort scores throughout the building. She undertook to visit areas reporting problems with building noise control in order to understand where occupants were finding building noise disruptive to their tasks. This on-site research (data interpretation) enabled her to decide where she wanted to make adjustments to the building systems, and where she would recommend to tenants that persons carrying out that particular kind of work be moved to quieter areas of the office. Her short-term actions included contacting the building systems contractor for air quality testing and HVAC system analysis. Her long-term plan included replacing the building's lighting system with a more flexible system that responds better to tenants' changing needs.

The completion of the pilot study enabled Comre Co.'s senior management to determine who on their staff to involve in a larger scale and more systematic implementation of an occupant feedback program. Company directors felt it was a major and somewhat risky step for a property management firm to involve itself so intimately with its tenants, providing advice on their accommodation and even sharing in their expansion or downsizing plans. However, marketing personnel felt that increasing property and facilities management outsourcing opportunities offered Comre Co. an opportunity to acquire more clients for property management services, and the board felt that occupant feedback was a tool to prepare them for entering this market with more confidence. The next stage of their BIU implementation plan was to develop a normative database for a series of buildings occupied by a single, government tenant in a large urban area. For this effort, the building administrators of the buildings involved were convened, as well as the staff members on the tenant's side responsible for facilities management and space planning. An individual from Comre Co.'s Head Office was assigned to the process, to direct the work, convene and run the meetings, and create liaisons between field staff and the head office.

BIU data, collected from the occupants of these buildings, were organized into a database from which BIU norms were calculated. The BIU profiles of each building were compared to Comre Co.'s norms as well as to the BIU norms: for the first time tenants were able to compare, systematically, the environmental quality of their offices to industry at large or to each other. As the age, size, and style of each building varied considerably, meetings to review and interpret results were convened in each building so that tenant representatives could be involved in each case. An action plan was drawn up for each building which set priorities on how Comre Co. would follow up on the information. Short-term interventions were identified; longer-term interventions were further analyzed for resources required and approvals needed. Company staff and the tenants' own managers drew up the action plan in concert and shared responsibility for implementing it.

As a result of this process, some of Comre's corporate tenants entering lease renewal negotiations requested similar strategic input. One tenant that participated in a BIU survey used the results to create an action plan for resources to enable them to make improvements themselves. The action plan was worked out with the advice and support of the Comre team, and the tenant was enabled to reach agreement with Comre over the changes and services Comre would provide with the new lease agreement. Tenant representatives even used the opportunity to share some of their company's long-term business plans: two offices were to be merged and space had to be reorganized to improve efficiency; a third was to be dissolved, with several employees being let go. As the departments reconfigured and resized to respond to market changes, so Comre Co. was in a position to collaborate with its tenants on strategic improvements to their O-A relationships. The relationship of mutual trust and concern that developed meant *both* that the tenant did not consider relocating to a new property *and* that Comre Co. could keep its rents at the top end of the market because of the value-added services it was offering. Not only were all upcoming leases renewed by tenants, but tenants in other buildings heard about the pilot project and asked to be involved. Comre Co.'s reputation as being more than just another property management firm is growing.

Impact of the Transition on Organizational Learning

With each tenant, the building administrator took care to communicate back to occupants the survey findings that generated action decisions. This information was later followed up with bulletins that identified the actions that had been and would be taken to improve their work environment. Some tenant representatives joined their communications with Comre's, and others distributed memos of their own containing a summary of their interventions as well as advice to occupants on better ways to manage their space.

Comre Co. surveyed the same buildings again after a year to measure the impact of the process. The data showed improvements in functional comfort levels in nearly every building. Overall, four of the company's BIU norms had gone up and three had stayed the same. None had come down. The norms were adjusted accordingly for 1993, and the action plan was brought up to date. Other issues were now a priority for the building administrator, and some of the previously identified problems had disappeared. The process was simpler the second time around and had the additional advantage of providing proof to head office that the actions taken had paid off. The administrators noted that responding to the occupants' priorities had not increased their expenditures overall, but had caused them to assign their resources differently than they had planned. The costs of collecting and analyzing the occupant feedback data had been more than offset by the increased efficiency of building management staff resulting from the reduced number of occupant complaints. Additional advantages included a more satisfied group of people, better deployment of their own resources, and a higher level of commitment among their own staff.

The company's progress into this new role was not unmarked by conflict. Members of the company's Board of Directors feared that conflicts of interest could arise when, as landlords, they provided advice on their tenants' business operations. Others were concerned about Comre's liability, not only from becoming involved in tenants' business strategy, but also from advising tenants on better ways of planning and designing their own space. Some of Comre Co.'s staff had to be retrained and—reluctantly—prepared for more participation in their tenants' decision-making. Many building administrators and their staffs had adapted to the role of the "good service-provider", and did not see their accommodation expertise as a corporate asset to be mined. Many resisted greater involvement with tenants, feeling more comfortable and expert in the behind-the-scenes role of the property manager. Comre executives appointed a strategic planning team to design and implement the changes that were necessary in their company and gave them the power to evaluate strategically how effective their changes were in terms of increasing their competitive advantage. Any evidence that the risk was not paying off would have been communicated to business strategists immediately.

With their investments in new procedures, new norms and expectations, and retraining employees paying off through higher rents, longer leases, and reduced turnover of tenants, Comre decided to advertise their value-added services; they defined their market niche as supplying strategic accommodation advice to companies, especially small companies with little or no in-house real estate expertise. In the next stage of their business plan, Comre Co. undertook surveys in another city in which its buildings were occupied by a variety of different tenants. The number of possible uses of occupant feedback increased. Feedback from those whose leases were shortly to be renegotiated was directed to the space planning and design group as well as to the leasing staff. Together, they worked out a series of responses to the findings that equipped them for a strong position in the upcoming lease negotiations. Not only could they offer more and better space if it were needed, but they could also offer ideas for more efficient use of office-space and ways of improving its function as a tool for work. In this way Comre Co. showed its tenants ways of making their expenditures on space function as investments in the improved performance of their employees.

DEVISING CORPORATE ACCOMMODATION STRATEGY

Table 9.1 summarizes the stages of a strategic approach to accommodation planning for the three companies, based on the occupant feedback model presented in Chapter 4. This is the model whereby a company collects occupant feedback, uses it to formulate its accommodation strategy, and allocates resources to creating a useful space planning and communication tool. Although carried out quite differently in each of the case studies presented, the five stages identified in Chapter 4—

Table 9.1   5-Stage Summary of BIU Assessments for Strategic Accommodation Planning

Stage	International Co.	Telecom Co.	Comre Co.
1. Collecting Information *(and selection of buildings for surveys)*	All HQ buildings surveyed, resulting in amassing enough data to form own database, but generating too much information for facilities groups to respond to.	Two leased buildings selected on an experimental basis, followed by two owned buildings, followed by a 10% sample from all buildings. Information acquisition was therefore staged.	Pilot project showed how occupancy feedback from tenants could work in one building; several buildings occupied by same tenant selected for next stage.
2. Analysis and Interpretation *(and communication with occupants)*	Provided departmental managers with results. Staff sometimes sent memos out to occupants after delay. On some occasions, nothing was communicated.	Kept results within building services departments, but held focus groups with occupant representatives and sent memos out to occupants.	Tenant representatives heavily involved; both they and Comre staff communicated follow-up actions to occupants.
3. Action and Follow-up *(and development of recommendations)*	Sessions with occupant representatives to present and discuss survey results. Efforts made to get line managers involved.	Detailed walkthrough with building managers and, later, with senior staff and occupant representatives. Follow-up actions communicated through publications and newsletters.	Planning sessions with occupant representatives to develop accommodation strategy and negotiate lease conditions
4. and 5. Communication and Negotiation	Actions carried out as a result of survey findings not announced to occupants or to colleagues in other facilities groups. This reduced impact of intervention.	Occupant representatives involved in survey process, in interpreting results, and devising follow-up action encouraged to communicate actions to other employees.	Comre staff provided expertise to assist tenants in business planning, space forecasting and more efficient space use. Tenant representatives, intervened to improve employees' work environment.
6. Integration into the Organization	Individuals undertook follow-up actions they deemed necessary; a physical needs assessment of each building was initiated as a basis for action planning for each building.	Initiated a customer satisfaction survey in all buildings as part of its total quality program. Devised a strategy for communicating results to individual building management teams.	Used feedback to add value to services and increase market share. Systematized approach to occupants' feedback so that service could be offered in all buildings, on tenants' requests.

along with a sixth step of integration into the organization—are integral to the process in each case.

Based on the summary of the key differences among the companies in how they implemented the surveys and how they processed the results and follow-up actions, it is clear that considerable single-loop learning has taken place in Comre Co. Its managers learned to carry out occupant feedback studies in order to increase the company's competitive advantage. To accommodate these new activities, Comre Co. has had to change its norms and attitudes to some degree, and has had to resolve certain conflicts. To the degree that it has been successful in doing so, a certain amount of double-loop learning has occurred, but more may eventually be necessary as conflicts arise between the old and the new way of doing things.

International Co. had more experience with feedback from its buildings before starting with BIU Assessment and as a result in part of its sophistication and in part of its poor intergroup communication mechanisms, did not prepare its facilities staff for stages in the strategic feedback process other than data collection, analysis, and interpretation. Consequently it did not evolve a new way of doing things, but it was able to apply some the acquired information to solving short-term problems and conflicts within the confines of its current role. Telecom Co. moved more cautiously than either of the other two companies, but derived considerable organizational learning from the experience. As a large and well-established corporation adjusting to a climate of change and competition, it is moving slowly towards radical change, not least among which is the role of the facilities department in developing and implementing business strategy.

All these companies, in different ways, understand that a good manager-occupant relationship does more to reduce building operating costs than any other single factor. A good relationship and good communication is the only way to encourage occupants to take responsibility for their building—a key element of their acceptance, comfort, and effectiveness in the building. If occupants feel they understand how their space works and can make it work for them, then they will communicate with their facilities team to make it do so. And it is at this point that the office environment can function as a true tool for employees to get their work done. The potential of the strategic approach to accommodation planning is explored in detail in the next chapter, as we review some of the opportunities and difficulties of mounting a systematic approach to occupancy feedback in an organization. The communication potential and how to realize it requires strategic planning, as well as tactics for conflict resolution. The related issues of employee empowerment, total quality management, and environmental negotiation are key procedural concepts for the implementation of a successful accommodation strategy.

NOTES AND REFERENCES

1. The case studies reported in this chapter are hybrid cases drawn from approximately seven applications of Building-In-Use Assessment to organizations.
2. C. Argyris, and D. Schön, *Organizational Learning: A Theory of Action Perspective* (Reading, Ma. Addison-Wesley, 1978).
3. Argyris and Schön, *Organizational Learning*, p. 18
4. Argyris and Schön, *Organizational Learning*, p. 24.
5. Senge, The Fifth Discipline, p. 114.
6. C. Argyris, and D. Schön, *Theory in Practice*, (San Francisco, Ca: Jossey-Bass, 1979), chapter 3.

CHAPTER 10

THE POLITICS OF OCCUPANTS' FEEDBACK: ISSUES IN IMPLEMENTATION

"[Frank Lloyd] Wright was a master of strategic work. He correctly understood the important social–spatial issues at stake. He understood that they would be decided by the voters."

Glen Robert Lym, *A Psychology of Building*

In previous chapters, opportunities for feedback from occupants about their workspace have been explored in order to demonstrate that companies can make better use of their real estate and space costs through creating a better fit between the tasks people do at work and the physical environment in which they carry the tasks out. It has also been possible to outline, in some detail, a system for acquiring and using feedback from building users in order to make the kinds of changes necessary to improve work effectiveness and productivity. Acknowledging the rationality of this argument and the sophistication of BIU Assessment as an approach, still, every company has norms, attitudes, values, and ways of doing things that create a highly political environment in which to carry out this sort of activity. Therefore, good sense and rationality alone will not guarantee the success of a BIU-type intervention: understanding the politics of the organization is necessary in order to make it work.

The politics of a corporate environment can be a determining factor in the success or failure of an occupant feedback intervention, and this factor exists in small companies or small-scale applications of BIU Assessment as well as in large-scale examples. In this chapter, a summary of these types of opportunities is described, followed by a discussion of strategies for improving communication with occupants. The chapter closes with a discussion of who should initiate an occupant feedback system, and how this responsiblility affects their role in the organization.

APPLYING OCCUPANT FEEDBACK

Experiences with BIU Assessment—some of which have been described in previous chapters—have shown some seven ways in which occupant feedback has been and can be used to improve the O–A relationship. Each example demonstrates an approach to solving a particular problem for the organization. The seven types of occupant feedback application are:

- comparing before with after
- predesign programming
- planning, budgeting and priority-setting
- lease negotiations
- planning alternative workspace
- information exchange
- defusing territorial conflict

Comparing Before With After

Surveying occupants of a space before and after a physical change indicates to designers and managers how well they have succeeded in effecting improvements. This feedback is especially valuable when the principal goal of workspace planning is increasing functional comfort as opposed to simply supplying new furniture or making a move. In the example cited in Chapter 5, occupants were moved into a new space with new furniture, lighting, and layouts. Feedback was collected after one month and again after three months to examine the impact of such a move over time: the managers wanted to know if the strong positive response in users' comfort ratings held up as occupants became accustomed to a new environment.

The results showed evidence of a "Hawthorne effect": functional comfort ratings were considerably higher immediately after the move than they were after two months in the new environment, reflecting the glow of excitement in the move to new space. After two months, most scores were still significantly above the BIU normative scores, however, showing that there was a long-term increase in occupants' functional comfort.

This case shows that in order not to be confused by before and after comparisons, the Hawthorne effect must be taken into account, and neither dismissed nor relied on too heavily. Most office workers respond positively to any change made to the work environment on their behalf, and therefore comfort ratings in most *after* situations are higher than in *before* situations unless there is something horribly wrong. As the example given in Chapter 4 demonstrates, ratings of air quality and thermal comfort can go up after painting walls, recarpeting floors, and installing sealed windows in a formerly windowless space; yet none of these actions directly affected ventilation and temperature control. While the facilities planners may not have expected this feedback, it does not mean that the positive impact of the change should be dismissed: the Hawthorne effect is not an illusion. Environmental changes, even small ones, make people feel better, and this effect is integral to the concept of occupancy feedback. Part of the reason feedback is sought from users, after all, is because it makes them feel their opinions are important, which in turn augments their satisfaction with their environment. Using, rather than dismissing, the Hawthorne effect inspires a strategy of implementing a series of small environmental changes based on user feedback and ensuring that occupants are aware of managers' and planners' actions. Over the long term, such a strategy will cause comfort ratings to rise more consistently than one or two single large changes to the office environment about which occupants are not fully informed.

Predesign Programming

Design processes for new space, whatever the scale of the project, usually include a period of information collection about how people will use the space to be designed. This procedure is sometimes cursory, and often entrusted by naïve clients to design professionals with little experience or interest in pre-design programming. Much of the information needed for design decisions rests on the projected number, type, and activities of the future users of the space. Surveys are often used both to collect this information, and to provide designers with qualitative feedback about occupants' preferences and priorities regarding future space.

Usually, programming gives occupants little say in the final outcome, confining them to an information-provider role in the early stages of design. The designers use occupants' information to help make design decisions, and the proposed designs are then presented to chief executives or a senior manager for review and approval. An occupant feedback system that employs not only a user survey, data interpretation, communication and negotiation over follow-up action, is not usually part of conventional pre-design programming. As a result, an important opportunity is lost: that of having occupants involved in making decisions about their workspace design that can eventually make their work more efficient.

If some form of occupant feedback is substituted for or added to conventional programming, then the predesign process can be both simplified and speeded up. Moreover, architectural design would in such cases not merely be confirming the traditional top-down approach to space decisions in which architects and CEO's work together, and users resign themselves to whatever they are given. The example provided in Chapter 4, in which a BIU Assessment was used in lieu of conventional programming for the redesign of a law firm's offices, went part way toward occupant feedback objectives by engaging users in a change-generating process, and integrating their feedback into design decisions for the new space.

Planning, Budgeting, and Priority-setting

Occupancy feedback is a cost-effective tool to manage budgetary priorities, and to help plan the cycle of repairs and renovations that is necessary in all buildings. One of the examples in Chapter 5 described how feedback from occupants was used to find out how users experienced an older, somewhat uncomfortable building about which serious indoor air quality complaints had been received in the past. The feedback enabled managers to set priorities for the actions they wanted to take and to improve the building in response to occupants' own priorities for change. Rather than spending the whole budget on HVAC upgrades— as they had planned because of the number of called-in complaints about temperature and humidity—they elected to respond to feedback on occupants' concern with tobacco smoke leaking from the smoking rooms on each floor into the office space. The facilities team installed a dedicated exhaust fan in each smoking room, to draw smoke directly outside the building, and then used the leftover dollars to paint the walls and to purchase indoor plants for the offices. The net impact of these actions was a far more substantial increase in functional comfort than would have resulted from their original plan.

Another building turned out *not* to be rated as uncomfortably as managers expected when the occupants were surveyed. Managers responded to the findings by locating the worst-rated areas and following-up with instrument measurements of thermal comfort and ventilation conditions in these targeted areas, only. By initiating repairs only in those areas of the building where problems were found, they both saved money and increased employee satisfaction with their services by responding promptly to users' feedback. They also learned that relying exclusively on service calls from occupants gave them a biased and imprecise definition of the problem needing solving.

The systematic and proactive nature of an occupancy feedback process gives planners diagnostic information about the functional comfort of the work environment which they can use for establishing priorities on future changes. This strategy is an improvement over spending money reactively to fix problems identified by individual building users, which may or may not improve employees' functional comfort.

Lease Negotiations

Companies who are tenants in modern office buildings use BIU Assessment to collect data that can be applied to lease negotiations with their landlords and property management firms, while at the same time showing concern for their employees. Since employees' functional comfort affects their productivity and, therefore, the profitability of the organization, tenants have a right to ensure that the office environment they negotiate from their landlord does not in any way reduce employees' effectiveness. Typically, a property owner provides a base building and building systems, with the tenant "building out" office space and designing a layout to accommodate the furniture. Tenants can only hold the landlord responsible for those base building attributes that affect functional comfort, such as air handling systems, heating and cooling, and lighting, while taking responsibility themselves for furniture comfort, privacy, and noise control. However, the lines between these two spheres of responsibility blur when landlords provide design services or build out the space according to tenant specifications as part of tenant improvements, and when tenants depend on the property owners for advice concerning suitable office design and space layout in their building.

As the Comre example chapter 9 shows, property management firms that take a proactive approach to meeting the needs of their tenants seek out employees' functional comfort criteria in order to provide better space. However, more typically, landlords are reactive: they prefer to wait until tenants complain or demand change. Tenants place themselves in a stronger negotiating position than landlords by invoking the functional comfort needs of their employees as the basis for acceptable lease terms. In a typical example, occupants who were surveyed by their own managers rated air quality low in a building where the landlord had not been able to find specific ventilation problems. Interpretation by the tenant showed that an increasing number of staff were working late in the evening and coming in on weekends, and the discomfort they were experiencing was because the building's air handling systems were switched off after usual working hours. According to the lease, the tenant was charged a significant sum to receive longer hours of HVAC operation. When the lease was renegotiated, one of tenant's priorities was longer hours of HVAC operation at no extra charge—and they got it.

Planning Alternative Workspace

Functional comfort criteria are critical in planning office alternatives for employees. If an organization wants to determine whether members of a workgroup should be telecommuting, working in satellite offices, sharing workspace, or accommodated in a Universal Plan, it must first understand employees' task requirements, their way of working, and their environmental quality criteria.

One large company targeted six workgroups spread over two floors of its office building for alternative workspace, applying criteria such as size of group (medium), type of work (sales and customer support), and willingness of their managers to participate. BIU Assessment results from each of the groups allowed the company to establish a short list of most suitable candidates in terms of the workplace alternatives available. The results of each survey were then presented to each group, and a participatory process was established that encouraged workers not only to replan their work so that it could be carried out on a decentralized basis (technology support, necessary information, etc.), but also to select and define the most appropriate environmental alternatives for each group.

One interesting outcome of this process was a reprioritizing of the seven BIU dimensions for users when functional comfort criteria were applied to the planning of alternative workspace. Air quality, for example, while important in the company's large office building, gave way in importance to lighting comfort and noise control in satellite offices; and spatial comfort, while important when everyone had his or her own desk, dropped in importance when workgroups discussed shared workspace, group meeting areas, and equipment access.

Information Exchange

Occupants' complaints are often more attributable to a poor relationship between facilities managers and building occupants than to technical problems with the building. If employees do not understand how a building's temperature control system works, or how to adjust the ergonomic furniture, or why the lights dim when the sun comes out, they resent building management staff for failing to provide them with a functionally comfortable environment. Better informed employees are more functionally comfortable, and have a better attitude towards building managers. For their part, in requesting occupants' feedback, managers open up an opportunity to receive information about how occupants actually use the building and which environmental elements have the most impact on their functional comfort.

After a BIU Assessment to help occupants understand the nature and extent of a proposed building renovation, one facilities team started a facilities management newsletter which was sent out to employees on a regular basis to keep communications open. In another situation, managers waited to start a planned upgrade to the ventilation systems until employees had provided feedback on their air quality and thermal comfort. This tactic enabled them to demonstrate their responsiveness when the low comfort ratings they expected were received. Another manager wanted employees surveyed before and after a move to new premises, not because he wanted to measure the increase in comfort and satisfaction, but because he wanted the opportunity to "sell" the new space to employees by demonstrating how their criteria had been addressed. Managers have also used occupancy feedback to demonstrate to nonfacilities staff and workgroup managers how their people feel about their space. Such data involve non-facilities personnel in issues that affect building use and assist them in decisions that pertain to space quality.

Defusing Territorial Conflict

Resistance to change in organizations is often expressed in hunkering down, us-against-the-world behaviors that are manifested in people's use of space. People become defiantly territorial, pushing out and closing their boundaries or boundary markers, personalizing and labeling their space and things, and aggressively laying claim to furniture and storage space. Unfortunately, the history of the office as a personally assigned symbol of status and role—and not just space to work—has encouraged people to use space and spatial imagery to express their personal selves.

Of all the elements of an office space, people hold on most tightly to these: walls and partitions ("for privacy"), storage space ("for all my files"), and square feet ("I need more space"). And even if impending change is not threatening, people never feel they have enough of these three elements, and they always blame managers for not providing more.

The notion of office space functioning as a tool for work runs counter to this emotional stake in space; unending frustration is generated in both employees and managers when the managers are pushing to streamline and "functionalize" space—open it up, create team space, and share tools and information—while the employees are shoring up their defenses and digging in to hold on to their territory. A BIU Assessment carried out in a building housing laboratories and offices for scientific research helped to bring these differences to the surface and to start negotiations to resolve conflicts between researchers about lab space. By focusing building users on the strengths and weaknesses of their workspace rather than on the amount of space they occupied, they were able to begin to make trade-offs based on the relative merits of different locations in the building rather than simply claiming additional square feet in order to add equipment or storage to their labs.

Are office workers as amenable to reasonable alternatives as research scientists in laboratories? Often more traditionalist in their attitudes towards work, office workers also have less of a stake in their workspace than researchers have in laboratories and, therefore, might be persuaded to focus on functional comfort issues rather than on walls, storage, and square feet. Managers who have been successful in bringing about this change in attitude have taught employees to dis-aggregate their tasks into all the myriad small activities they carry out during a typical work day, and then to examine the functional comfort requirements for each task. Employees who have followed this guide have been able to see and to understand that using space as tool for work is a more cost-effective and, in fact more natural than quasi-ownership of a permanent, fixed, assigned workspace with a certain limited number of attributes, which can only be improved by having more of those same attributes.

LEGAL AND POLITICAL ISSUES

In summary, occupant feedback can be applied to a range of decisions. These include:

- Immediately adjusting the building to solve problems and improve people's comfort—for example, tracing the source of reported odors and sealing the leak.

- Establishing how much is being spent on maintaining certain levels of quality—for example, if high spatial comfort scores are received, could less have been spent on the furniture system? If the air quality score is low, should more be spent on operating the HVAC system?

- Using occupant feedback as a training tool for facilities managers— for example, the results of a survey could be turned over to a facilities team who are then asked for a long-term plan that responds to the priorities expressed by occupants.

- Focusing long-term strategic planning on functional elements of the building, such as distance from home and travel and access requirements—as well as flexibility and adaptability of interior space, and effiency and adaptability of lighting and ventilation systems.

- Determining criteria for new space—for example, if a move is planned. Explicit criteria will improve the company's ability to find the right building for itself, and also will reduce the scope and time needed for the search.

- Teaching employees to think about a new way of working—for example, using functional comfort analysis and workspace planning to help people rethink how they do work to generate more effective and efficient procedures.

The value of an occupants' feedback system is that it is a tool for monitoring, measuring, and managing the O–A relationship. Carried out at regular intervals, systematically collected feedback from occupants serves as an early warning system to indicate shifts in functional comfort and discomfort levels before these materialize into larger and more expensive problems. The feedback process is iterative as the data can be updated by new surveys, stored, and accessed to serve a wide range of corporate interests.

Some managers have expressed concern about the legal obligations inherent in an occupancy feedback initiative. For example, does questioning people about their workplace comfort obligate the questioners to act on the information they receive? And can they be sued if information is acquired through a survey and then not pursued? There have certainly been numerous cases of lawsuits initiated by building occupants who felt their health or safety was threatened by inadequate ventilation (sick building syndrome), by uncomfortable computer furniture (repetitive strain injury), or by other endangering conditions. In these cases, the issue of whether or not a survey had been carried out was inconsequential in the face of medical testimony, evidence of negligence, and other considerations.

Of more concern is the use of occupants' feedback to bolster one side or the other of a lawsuit that has already been launched. This scenario occurred in the Boston building case cited at the beginning of Chapter 1, in which employees were encouraged to report illnesses and symptoms in order to support the tenants' case against the building owners. Initiating surveys of building users, then, is alone unlikely to engender legal difficulties, especially if information about the scope, purpose, and eventual use of survey results is clearly provided to all. However, this is not to say that participating parties in high stakes situations might not make unscrupulous uses of survey data.

Politically, seeking feedback from occupants of a building is also risky. In a large organization, surveying users about their workplace can pose a challenge to established channels of information flow: building managers relate in new ways to building users; senior executives relate in new ways to building managers. To successfully negotiate the rocky political trail of user feedback requires careful management of information, and a willingness to open up accommodation decision-making to employee scrutiny. Some managers feel threatened by the idea of asking occupants to assess their own work environment. For them, doing so means giving away power to ordinary clerical and technical workers, an idea that dismays many in spite of recent corporate empowerment philosophies. For others, it means having their job performance evaluated: not something they necessarily seek from coworkers and clients. For most, asking users to assess their work environment is indistinguishable from asking them for a "wish list", implying a promise to meet occupants wishes and needs—a promise that facilities staff feel cannot be kept because of their limited resources and the technological constraints of their buildings.

The distinction between asking building users to make a finite number of environmental judgments in response to questions posed by managers—called environmental assessment—and asking building users on an open-ended basis whether or not they are satisfied with services, and what is wrong or missing—which is the basis of satisfaction evaluation—is fundamental to a solid understanding of the occupant feedback process. Occupants who are told that their feedback is to be used to help provide a better work environment see the process as a way of collecting data on human perceptions of environmental conditions. If they are told that their feedback is sought regarding the shortcomings of their work environment, then they see the process as an opportunity to list everything they ever found wanting in an office, including better furniture, incandescent lamps, more filing cabinets, and, inevitably, more space. This type of response overwhelms managers and gives them the feeling that they have failed before they started. However, feedback from a structured assessment can be applied selectively to make improvements that are affordable and to respond in various ways to occupants' functional discomfort. As pointed out in Chapter 3, increasing users' satisfaction with building services is not the same as using occupant feedback for strategic planning of the O–A relationship, although both have their place.

IMPROVING COMMUNICATION

In organizations that make extensive use of surveys to determine employees' job satisfaction, stress level, and other "attitude" issues, as well as satisfaction with building-related and other services, senior executives often see occupant feedback surveys as an "employee empowerment" technique.[1] But while managers may favor such an approach, facilities staff fear that employee feedback makes their building even more difficult to manage. They are also aware of the struggle they have to get budgetary resources for the building improvements that they fear will inevitably be demanded by empowered employees. At other times, the opposite is also true: corporate executives may withhold approval of an occupant feedback initiative in spite of a perceived need for it at lower levels of the organization, because they are the ones who fear not just empowerment but also the possible dollar implications of asking employees for their opinions about the work environment. In some organizations, employees themselves may not favor an occupant feedback initiative because they feel their time at work should be spent on more important activities, and that space and space-related issues ought to be someone else's headache. One cannot predict, however, when a space-related problem will unpredictably arise, consuming untold hours of people's time to resolve and adding unforeseen sums to a building's operating budget.

Regardless of who takes responsibility for the O–A relationship, an occupant feedback initiative is a major opportunity to empower employees, to instruct them in judging their own functional comfort, to open up dialogue between building occupants and managers, and to encourage facilities staff to move beyond their customary reactive and potentially adversarial relationship with users. As facilities planning becomes more integrated with the business mission of the modern organization, so the potential for effective communication and negotiation by users and managers of space is growing. In fact, the style, amount and structure of user-manager communication is likely to change still further as the role of facilities management financing changes according to the sequence outlined by the MIT team discussed in Chapter 2.[2] In Table 10.1, each strategy that the report describes is characterized by variation in the style both of occupant feedback and of user-manager communication about the work environment. Feedback from occupants about the functional comfort of their space becomes more central to real estate operations as facilities management becomes more successful as a profit center, as it competes with the market in providing space, and as it becomes more integrated with the business activities of the organization. But what about problem situations in older buildings, where there is little hope of improvement, with or without feedback?

Table 10.1   Impact on Building Manager—Occupant Relations of Real Estate Financial Coping Strategies

Real Estate Coping Strategies	Impact on Managers and Occupants
1. Engineering approach	Managers are service providers, quality of service is good if money no object; managers are "behind the scenes," no communication with occupants, other than to get "wish lists" and service calls.
2. Cost minimization	Managers police the provision of accommodation, impose standards; as a result, adversarial relationship can develop with occupants, sometimes an effort is made to survey users' needs to help make more informed cost-cutting decisions.
3. Market cost and usage standards	Managers start to see themselves as eventually competing in the market, so they improve customer service, research customer satisfaction, and develop individual manager-customer relationships.
4. Market design approach	Managers initiate environmental negotiation with occupants; customer satisfaction surveys are replaced or augmented by more functional information on occupancy needs and the requirements of work.
5. Business strategy approach	Occupants' feedback is a routine part of accommodation decision-making; functional comfort is a requirement of all environmental planning; processes are in place for negotiation of environmental solutions, communication is ongoing.

Many office workers go every workday for years to a building they do not like and in which they are not comfortable. Sometimese the desks are open and exposed ("bull pen" lay out—see Figure 10.1), and sometimes they are partitioned off in individual cells (see Figure 10.2). Typically, this building is older, dating from the seventies or perhaps the sixties. Modern office technology and modern attitudes towards office workers were not part of what shaped this building, and the environment is unkind and depressing. Inside, it may have dark brown carpets and yellow lights; or perhaps it has a bright green or purple carpet whose color dominates occupants' field of vision. It has large sealed windows that let in lots of light and heat, but no air, and need to be covered up on sunny days or where computer screens are in use. It has small gloomy washrooms with nowhere to place things, and an insufficient number of small, slow elevators.

Because there are more people and equipment in the building than was ever planned for, the building gets stuffy, is often too warm, and feels airless, especially in the afternoons. The occupants do not understand HVAC technology, but they have heard of sick building

Figure 10.1. Noise from people and equipment, bright lights, and no privacy make this an uncomfortable workspace.

Figure 10.2. A forest of high partitions looks cluttered, blocks airflow, and cuts off people from one another.

syndrome. Eventually, there is an odor or other evidence that the building's air is inadequate. Flu and colds are easily transmitted; there may be other illnesses. There are certainly headaches and eyestrain. People's chronic discomfort in this building generates an endless series of problems for the facilities staff, often culminating in a sort of sick building hysteria. People complain without hope of their problems being solved, occupant morale is low, and demands on facilities staff are unremitting. There is little time for staff to plan or allocate resources to long-term improvements, because so much effort and money has to go into crisis management. It is difficult to keep such buildings clean, and employees do not take care of either the building or the furniture. People's discomfort translates into lighting problems, ergonomic problems, air quality problems, and, sometimes, an organized protest or even a walkout. The ongoing disgruntlement of employees who feel they are being treated less than well by senior management can generate adversarial union–management relations, as well as bad feeling towards facilities staff, space planners, and related building professionals.

Senior management can try to ignore these problems and state that in times of cutbacks and unemployment people should just be glad to have their jobs. At other times, senior management tends to view these situations as tinderboxes. They handle complaints that emanate from such buildings with kid gloves, and often large amounts of money are spent on accommodating employees' apparent desires, such as more parking spaces, a new cafeteria, or a fitness center. What they cannot do is finance a move into a new building, either because they are locked into a long-term lease or because they own the building and cannot afford to vacate or upgrade it.

In view of the high level of discomfort and resentment among the occupants of such buildings, is occupant feedback still a useful tool? For those involved in the day-to-day management of such buildings, the notion of asking occupants for their judgments of the building's environment is little short of insanity. But, in fact, opening up a dialogue between users and managers of a building can serve a therapeutic purpose in situations like these. Being constrained by the type and content of feedback requiring them to do no more than provide their perceptions of certain environmental conditions, occupants are in a position to respond responsibly to the request that has been made of them. They are not taking a public stand, or trying to influence management opinion: they have been asked to participate in a process of information-gathering. By responding to this initiative, occupants find that not all of the building is bad. There are, in fact, when they stop to think about it, certain aspects that are quite comfortable and that they like, perhaps its location. More awareness of issues of environmental quality helps people tune in more precisely to how well they can get work done, often to find that more filing cabinets or a window view are not as necessary to them as a lower noise level or better lighting on their computer screens. They may as a result become less combative, less resigned, and more productive, because they see and can communicate to managers what they need to use their workspace as a tool.

One senior manager, knowing that air quality was in fact the one serious problem in his building, had already developed a strategy for implementing improvements to the air handling systems before the survey of occupants' functional comfort was carried out. When he heard back from occupants via the BIU survey that air quality was their number one problem, he was able to respond promptly by acknowledging that he had heard them, and by describing the renovations that were about to get under way. In handling the situation in this way, he was sure to get full credit for solving the problem, and employees felt their voices had been heard.

ENVIRONMENTAL EMPOWERMENT: Are We Ready?

Much of the dissatisfaction users express with their work environments has been attributed to lack of environmental control.[3] Most large office buildings traditionally supply light, heat, and air (or cooling) in ways that are outside the control or even the understanding of the individual user. Being unable to switch off lights or switch on heat (without disturbing coworkers) contributes to a level of frustration with the work environment that is exacerbated when the workplace is uncomfortable and presents problems. A common assumption is that if people were able to change thermostats, adjust their furniture, turn their air on and off, and dim their lights, both their satisfaction and their comfort would increase.

Another sort of control is provided by offering people a chance to participate in the creation of their work environment. This does not mean protracted and detailed participatory space planning processes which consume time and resources by tying up space planners in endless redesign of schemes and layouts. It means asking occupants for their input, acknowledging that input, managing it according to the resources and opportunities available, and then using that input and letting people know it has been used. Once changes have been effected, input from occupants is sought again. Occupants' input does not have to be addressed to the same way every time, and circumstances will be different owing to different opportunities and constraints, changes in the workgroups themselves, and perhaps even changes in the philosophy of the organization.

This process of environmental negotiation is key to the quality of the work environment. Through the feedback and negotiation process, occupants become aware and informed about their workspace and become trained in defining how best to meet functional comfort goals. In offering environmental empowerment to employees, managers of necessity presume organizational readiness for a broader employee empowerment philosophy. It is important to remember that environmental change is also an opportunity for organizational change, and also for reengineering work processes. Ultimately, an occupant feedback process generates a level of environmental empowerment among employees because of the control they gain over environmental decision-making. Once occupant feedback is accepted, it is apparent that there is no ultimate total quality workplace, there is only a total quality process that generates a comfortable workspace that helps people get their work done, and that lends itself to change and adaptability as required by occupants in accordance with their changing work requirements. The "negotiated solution" is more effective than the "perfect fit" in solving the problem of effective accommodation for work.

One company uses an occupancy feedback questionnaire two or three times per year to monitor whether they have progressed towards environmental quality goals. Integrated with a customer satisfaction survey on the services and performance of facilities staff, the functional comfort questionnaire results show FM staff how they can improve on their services to users as well as levels of users' functional comfort in the buildings they occupy. If the results show an improvement on the previous survey results, staff are rewarded with promotions and bonuses; if they do not, staff are not promoted even if they are due for a pay increase. In a company where facilities services are a profit center and must earn their revenues from client groups within the organization, the success of the environments they provide has a direct impact on their bottom line. Client groups are free to retain commercial property management firms or facilities teams from other companies if they are dissatisfied with FM services. This arrangement functions more effectively as an environmental control for clients than any number of knobs and whistles on their furniture. And the need for both clients and managers to protect their bottom lines is a stronger incentive to environmental negotiation than any amount of heartfelt urging by the company's total quality experts. Such an arrangement gives users control over the quality of their workspace in much the same way as reengineering work processes gives employees control over their tasks—through a lot of hard work.

The right kind of information, and an organizational environment prepared for full use of that information, means

- defining and articulating the goals and purpose of an occupant feedback initiative,

- defining strategies for user involvement, and structuring opportunities for employees' environmental empowerment,

- having an end result or product in mind based on communication and negotiation between the users and managers of space.

Unless these preparatory steps are taken, this information, like all the other rivers of information that flow through organizations, will flow nowhere. Decision-makers in government and corporations, overwhelmed by the quantities of information that are available to inform their decisions, tend to simplify the processing of the information they receive. They do this by only processing information they actually need for a specific situation, by trusting one or two sources that have special expertise needed for that problem. Occupant feedback, if well executed and well-managed, is more likely to reach busy strategists who have to determine whether or not to move into a new building, whether or not to buy a new line of furniture, or whether or not to support an investment in flat screen technology—all decisions with important monetary implications for the organization. In the next chapter, these decisions and others like them are placed in the context of strategic business planning. Given the current business climate, how can accommodation planning be integrated with business planning to add value to a company's products and services?

NOTES AND REFERENCES

1. Claudia Deutsch, "Asking Workers What They Think" *The New York Times*, 22 April, 1990, p. 29.
2. M. Joroff, M. Louargand, S. Lambert, R. Becker, Strategic Management of the Fifth Resource: Corporate Real Estate (Report of Phase One Corporate Real Estate 2000: Industrial Development Research Foundation, 1993) pp. 40–52.
3. V. Hartkopf, V. Loftness et al *Designing the office of the Future: The Japanese Approach to Tomorrow's Work Place* (New York: John Wiley, 1993) p. 175.

CHAPTER 11

Optimizing Occupancy: Strategic Planning of the Organization–Accommodation Relationship

"Learning organizations themselves may be a form of
leverage on the complex system of human endeavors.
Building learning organizations involves developing people
who learn to see as systems thinkers see, who develop their
own personal mastery, and who learn how to surface and
restructure mental models, collaboratively."

Peter Senge, *The Fifth Discipline*

ACCOMMODATION STRATEGY AND THE O–A RELATIONSHIP

Space is a strategic resource to businesses on two levels. One is the
physical deployment of its personnel in the accommodation the company provides for them: how much space it needs to accommodate its
people and their tasks. The other is the functional support the workspace provides to people engaged in specific, business-related tasks: the
work environment as a tool for work. Appropriate deployment in space

is cost-effective to the firm by minimizing its accommodation costs and optimizing the investment it makes in space. Functionally suitable workspace adds value by reducing downtime and increasing the efficiency of repetitive tasks. Space also provides both communication and separation opportunities to maximize individual and group creativity. Strategic decision making about accommodation incorporates both definitions of space as a resource; accommodation strategy is the topic of this chapter.

Take the example of a grocery chain planning to consolidate its bakery operations to increase market share. It has to decide on the relative merits of two bakeries. In determining whether to move the two operations under one roof or dispense with producing its own cookies and retain only one of the two operations, the firm examines the relative costs of operating the accommodation occupied at the two locations, as well as the performance of workers in the two locations in terms of their cost-effectiveness to the organization. The evident advantages of one of the facilities—a newer building, fewer workers, lower operating costs, a better worker-management relationship, and the relatively high costs of competitors' goods—will eventually cause the company to retain its baked goods operation and to merge its cookie-making operation in with it. Its decision-making strategy addresses accommodation issues along with business issues and incorporates assumptions about the value of good space to employee productivity.

Small successful firms that are growing have another sort of challenge. These companies outgrow their space rapidly and have to factor into their overhead the costs of finding new space, leasing or buying it, and moving into it. Such firms sometimes have to shrink down just as precipitously. User-environment issues are part of business planning as CEOs decide *when* a move is needed based on how many people are being crammed into a workspace or how dispersed people are; *where* to move based on employees' criteria for good workspace to perform work; and *how much* to spend on new space based on functional comfort priorities. Strategies to streamline this decision-making process so that it is shorter, and to inform the decisions so that they are made better, adds value to the products of such companies. Companies make better investments if they incorporate criteria affecting the quality of people's workspace as well as the quality of accommodation for the production of goods.

This being the case, how misguided, then, is the attitude of many public agencies towards space. Preferring not to attract adverse public reaction by spending public funds on their accommodation, many government departments in United States cities, towns, and states send their employees to work in space that is overcrowded, overheated, windowless, colorless, and filled with ancient furniture. Although the idea of accommodating public employees in luxury office space is unattractive to the taxpayer, this attitude does not admit the possibility that even government workers may perform their jobs better, have higher morale, and be more productive in space that functions as a tool for work.

Whether public or private sector, a company's accommodation strategy is designed to define the optimal relationship between people's work and their space. In the following section, a closer analysis of how accommodation issues can be incorporated into business planning is outlined.

PLANNING THE FUTURE
OF THE O–A RELATIONSHIP

The goal of corporate accommodation strategy is to optimize the O–A relationship. In view of the often typically poor state of that relationship, strategic accommodation planning needs to address at least three areas of quality improvement.

- *First, companies need to demand and expect more from their buildings.* The building industry is alert to changing client expectations and developers are prepared to research what goes into making a quality building other than a marble-finished atrium. Building experts, such as mechanical engineers, are capable of designing ventilating systems that provide better indoor air quality, consume less energy, and control thermal comfort more precisely, if they are required and expected to do so.

- *Second, all the professionals involved in the various stages of the building-occupant relationship need to share a belief in improving the quality of this relationship,* not just from the point of view of their own involvement in the process, but also from the point of view of other stages in the relationship and of other professional disciplines. Currently, building professionals tend to approach their work rather as car mechanics do. They work to solve the car's problem and have it run efficiently once back on the road; they do not examine the relationship between the car owner and the way the car is running. If leasing agents, architects, designers, and facility/property managers model themselves on those in the counseling professions rather than on auto-mechanics, the quality of the services they provide will improve. They will have to examine the context of the problem they are being asked to solve in order to know what to do and what services the client needs.

- *Third, planning and acquiring accommodation has to be part of the organization's business strategy,* and as such expresses a philosophy of space management that considers accommodation an investment, not just because of the real estate value of buildings, but because of the difference good accommodation can make to the effective operation of the organization.

Judicious use of functional comfort data that derive from BIU Assessment and other measurement systems helps companies move towards these objectives in the following way. First, knowing more about what makes people work effectively causes business owners and employers to demand and expect more from the buildings they build and rent. Second, the articulation of these demands causes related professionals providing services to be more responsible and aware in defining a good relationship between a company and its building. And third, in reconciling employees' work-related needs with the environments in which they perform their tasks, a more explicit link is forged between a company's operations and its accommodation.

There is a vast difference between the strategic role of capital budgeting decisions a generation ago and the strategic role of capital budgeting decisions for today and tomorrow, as described in Chapter 1. The investment in plant, machinery, and the space to accommodate machinery that characterized manufacturing has been overtaken in recent years by expenditures on—rather than investment in—office space, furniture systems, and office technology. But even as these words are written, companies are embarking on analysis of the more complex implications of alternative work environments. For as telecommunications and other companies are realizing, employees who are liberated by modern technology from space and time constraints do not require conventional office accommodation and equipment. The challenge for companies seeking to tailor their accommodation strategy for the next few decades is to devise a system of alternative work environments that create value and increase the company's competitive advantage. In other words, now and for the future, an organization's "physical environment"—which is the scale, location, technology, design, and functional comfort of its employees' accommodation—has its own role in the constellation of factors that constitute organizational effectiveness and the management of change.

Waterman et al. have identified seven "S-factors" that interact to create "an organization." These are:

- structure
- strategy

- skills
- staff
- style
- systems
- superordinate goals[1]

The authors stipulate that organizational change is not a matter of any single one of these, but is the result of a complex set of interactions among these items, all of which are linked to each other and, centrally, to superordinate goals, the "set of aspirations, often unwritten, that goes beyond the conventional formal statement of corporate objectives."[2]

In view of the increasing complexity of accommodation issues for organizations, and the intrinsic link between corporate goals, personnel, structure and systems, and accommodation, especially as firms move towards new ways of defining workspace, the addition of "space" to the 7-S framework (thus making it 8-S) goes to emphasize the key strategic role an organization's space plays in its ability to meet the challenges of new ways of doing business.

In identifying space as an item in the 8-S framework of the organization, it is necessary to define how organizational management should address space both to understand its role better and to adopt an advantageous stance relative to space in and for the organization. Accommodation would appear to fit best into corporate strategy in terms of what is sometimes called "functional strategy," the implications of which are spelled out next.

CORPORATE ACCOMMODATION AS A "STRATEGIC FUNCTIONAL UNIT"

Traditionally, accommodation planning for organizations has been hidden in human resources (satisfaction of employees), in technology (investment in hardware and real estate), or in financial strategy (real estate values)—three of several possible "Strategic Functional Units" defined by Hax and Majluf.[2] In distinguishing between Strategic Business Units (SBU) and Strategic Functional Units (SFU), these authors define a SBU as a "distinct group of products and services sold to a uniform set of customers facing a well-defined set of competitors".[4] Strategic Functional Units drive each area of "functional strategy", and they are:

- financial strategy
- human resources strategy

- technology strategy
- manufacturing strategy
- procurement strategy
- marketing strategy.[5]

Accommodation, like these other functional units, has major decisions associated with it that comprise accommodation strategy. The best example from the above list on which accommodation strategy can be modeled is "technology strategy"—an SFU "currently considered to be one of the central functions in achieving competitive advantage" and relevant both to "the dynamics of technological markets" and the "laborious and delicate process of internal management of technology"[6]. Drawing on other authors, Hax and Majluf list the following categories of strategic decision related to technology strategy:

1. Technology intelligence
2. Technology selection
3. Timing of new technology introduction
4. Modes of technology acquisition
5. Horizontal strategy of technology
6. Project selection, evaluation, resource allocation and control
7. Technology organization and managerial infrastructure.[7]

They go on to list measures of performance related to technology strategy. Using this list as the model for *accommodation strategy*, an equivalent list of major categories of strategic decisions linked to accommodation and measures of performance related to accommodation strategy, reads as follows

---

1. **MAJOR CATEGORIES OF STRATEGIC DECISIONS LINKED TO ACCOMMODATION**

- **Accommodation Intelligence**

Information is needed about space that is available in the geo–graphic location in which a company might locate its employees and about its costs. Knowledge is needed of what competitors spend on accommodating workers and the operating costs of owned or leased space; the existing real estate market and projected

---

(continued)

changes; the supply of space in suburban relative to urban locations and their relative accessibility; impact of market conditions on lease conditions and conventions; strategic locations relative to work force; quality differences in space available.

- **Accommodation Selection**

The appropriate combination of accommodation alternatives must be managed so that space resources are matched to the jobs of the organization. Some of the issues to recognize are selecting the accommodation needed for product and process innovation, assuring the congruency of accommodation options (including work-at-home alternatives) with the business life cycle and the desired business strategy, and assigning the appropriate priorities to functionally comfortable accommodation.

- **Appraisal of Accommodation**

Evaluation of the quality of the accommodation occupied by the organization is needed. This evaluation contributes to the proper allocation of space resources, the correction of functional comfort problems, decisions about the costs and benefits of workspace alternatives, and more effective lease negotiation and facility management.

- **Timing of New Accommodation**

A move or accommodation change signifies the opportunity for change and/or innovation in the organization. The company may develop internal expertise in acquiring and operating buildings, or it may resort to external sources. The decision is whether to lead or to lag behind competitors in innovations such as working at home, providing innovative building amenities, or shrinking office space. The risks and benefits of each strategy should be identified. The firm may elect to keep some but not all accommodation expertise inside the company, ensuring the congruency of the accommodation innovation with the generic business strategy.

- **Project Selection, Evaluation Resource Allocation, and Control**

The principal concern is the allocation of resources to support the desired accommodation strategy. Issues to be addressed are: criteria

(continued)

for resource allocation, evaluation of innovative accommodation strategies, and availability of loosely controlled funds to support and plan innovative workspace alternatives.

- **Accommodation Management and Managerial Infrastructure**

This item is oriented toward the organizational structure of the accommodation function and includes the identification of horizontal coordinating mechanisms needed to exploit the physical relationships existing among the various business units and the activities of the value chain. Issues include: centralized control over space or decentralized to business units, centralized space support services (facilities management) or decentralized to business units, development of career paths for facilities managers, use of lateral mechanisms to facilitate sharing of space resources, degree of involvement of top managers in accommodation decisions.

Continuing with the same model, where Hax and Majluf have identified measures of performance related to each one of the Strategic Functional Units, the following summary shows ways to evaluate a company's accommodation strategy.

2. MEASURES OF PERFORMANCE RELATED TO AN ACCOMMODATION STRATEGY[8]

- **Rate of Innovation**

Measure the rate of accommodation innovation by tracking the relationship between innovative products and processes and accommodation solutions.

- **Productivity**

As with any measure of productivity, measure improvements in terms of the ratio of the change in output to the change in input, i.e., the improvement in the performance of the employees divided by the incremental investment in innovative accommodation.

(continued)

- **Rate of Return on Investment**

Measure the profit generated by the amount of accommodation investment, not only in terms of changing real estate values, but also in terms of increased employee effectiveness.

- **Resources Allocated to Accommodation**

Monitor the level of expenditures being allocated to the accommodation of the various projects and businesses and at the level of the firm as a whole. Relate these expenditures to the number of new products introduced, of new patents obtained, or the percentage of sales derived from new products and services.

- **Monitor Functional Comfort**

Monitor the level of functional comfort of employees in various projects and businesses and at the level of the firm as a whole, to allocate accommodation resources effectively.

- **Other Appropriate Measurements**

Analyze the relationship between expenditures on accommodation and levels of functional comfort for various workgroups and business units.

This outline of strategic accommodation planning as a corporate activity is based on one model; others are possible. In each case, corporate norms and practices, in addition to available in-house expertise (the FM role and function), will guide selection of the most effective corporate approach to an accommodation strategy.

USING ACCOMMODATION TO ADD VALUE TO PRODUCTS AND SERVICES

Several approaches can be used to define how value is created by investment in accommodation, and how this increases a company's competitive advantage. Porter's concept of the "value chain" incorporates all the activities performed by a firm in a particular industry.[9] He asserts that "differences among competitors' value chains are a key source of competitive advantage".[10] The value chain consists of value activities which are the "building blocks" by which a firm creates a product or service that is valuable to its buyers. The objective of generic business strategy is to create value (or the amount customers buy) that exceeds the costs of production.

Porter distinguishes between a firm's primary activities and its support activities, both of which constitute the value activities of the firm. Primary activities are those involved in the physical creation of the product or service and its sale to the buyer; support activities "support the primary activities and each other by providing purchased inputs, technology, human resources, and various firm-wide functions."[11] They can be associated with specific primary activities, or with the entire value chain. Porter concludes:

> How each activity is performed combined with its economics will determine whether a firm is high or low cost relative to competitors. How each value activity is performed will also determine its contribution to buyer needs, and hence differentiation. [These differences] determine competitive advantage"[12]

The activity of occupying space clearly belongs in the category of support activities. Although not intrinsic to the production of the firm's product or service in the way that inbound logistics, outbound logistics and marketing and sales are in terms of its effect on competitive advantage, occupying space is nonetheless as critical a support activity as procurement, technology development, human resource management, and firm infrastructure (which includes legal affairs, planning, finance, accounting, and quality management).

Value activities are interdependent: the value chain is a system. Linkages between value activities show the impact of the way one activity is performed on the cost of another activity. *Optimization* and *coordination* are two ways in which linkages can lead to competitive advantage. For example, linkages between technology development and accommodation are critical in the definition of cost-effective workspace. The technology used by employees requires coordination with accommodation options in order to be used effectively: poorly planned or inappropriate space easily reduces the effectiveness of technology, whether it is up-to-the-minute communications technology in its relation to office space, or new medical and biomedical technology in its relation to research and hospital space.[13]

Using technology to liberate sales personnel from office-based work or to encourage professionals such as lawyers and accountants to work in their clients' offices, adds value to the firm's services by reducing the space requirements of personnel and improving the personnel's effectiveness. Firms that have invested in effective communications technology and in space-sharing furniture systems rather than in leasing or buying space therefore have a competitive advantage. The strategic importance of linkages such as these is often overlooked; Porter stresses the importance of acquiring information on linkages that will allow optimization and coordination to take place. The purpose of the value chain concept is to provide a framework for analysis of a Strategic Business Unit that will yield opportunities to increase value, and, thereby, competitive advantage. It is not until a company's accommodation is analyzed in this way that opportunities for increasing value as a result of improvements to its accommodation can specifically be identified.

In Chapter 1 we compared a bank which recuperated a nine million dollar expenditure on space and then went on to make money from successfully presenting its streamlined building to its investor clients, with another bank which made decisions about its accommodation outside the framework of its corporate strategy. One can predict which of the two will find most favor with shareholders in a few years time. A comparison of the two cases makes it clear that there is a cost to maintaining the separateness of the people that plan and manage space from the people who occupy and have feelings about the space and from the people that make corporate strategy decisions. The result of this separation is that the people who carry out the mainstream activities of the company cannot take responsibility for the quality of the workspace they occupy. Only when the people managing space are integrated with the people managing people, will the link between a better building relationship and increased effectiveness and productivity of the organization become part of the organization's way of looking at itself. In this case, the Bank of Boston clearly defined its accommodation decisions as a value activity of its business units.

REENGINEERING THE OFFICE AS A TOOL FOR WORK

Corporate attitudes towards space have traditionally varied according to the values and culture of the organization. In manufacturing companies, for example, office space is usually that which has been left over from space occupied by the manufacturing operation. In today's terms, this space is not often high quality space. For banks, managing the public's money means a chic and powerful image, but not particularly luxurious workspace for employees; it should be accessible but secure. Law firms typically select expensive space with well-appointed interior finishes that impress and awe their clients. Government agencies save high quality luxurious space for senior bureaucrats and elected officials, confining the large majority of civil or public servants to monotonous and poorly-planned interiors on the grounds that taxpayers are reluctant to see public funds spent on government workers' accommodation. Accounting firms have a traditional approach to space: expensive and well-appointed, but hierarchical. Partners and executives are accommodated in large, windowed enclosed offices with secretaries in workstations posted nearby. Junior clerks and accountants are accommodated in windowless interior rooms and small carrels that mark their inferior status in the organization.

These relationships all work after a fashion, but they are unexamined, and therefore less than optimal. As a result, opportunities for companies to derive more concrete benefit from their accommodation are foregone. Most companies see their workspace as overhead, and most employees see their workspace as demarcating personal territory.

However, in accepting people's functional comfort and a return on dollars invested in workspace as legitimate and realizable goals, an organization may come to see its accommodation as no more and no less than a tool for employees to get work done; the work environment designed to facilitate and expedite occupants' tasks in much the same way as the computer, the telephone, and even the humble pencil. In adopting the definition of workspace as a tool, questions of how much to spend on accommodation are answered in the functional comfort model by measurement of accommodation performance: investment in building improvements has to show measurable increases in the performance of the accommodation as a tool for work.

This model of the O–A relationship depends on the concept of workspace environment and users as a system, and feedback is a key element of its operation. According to Senge, feedback exists in all systems, balancing and/or reinforcing the system to protect its stability.[14] By structuring opportunities to elicit feedback in the user-environment system and by ensuring that the feedback is used to make O–A decisions, decision-makers are encouraged to understand and respect the system's own goals. In the event that these need to be changed (e.g., strategic changes to the O–A relationship), decision-makers' awareness of the system will ensure that they do not mistakenly try to substitute short-term goals to effect change (see the examples described in Chapter 9).

Feedback incorporated into the system as diagrammed in Figure 11.1 shows how the occupant feedback loop might work in a modern organization and indicates how the planning process has a momentum of its own. The planning process is informed by feedback from occupants, but not driven by it. The diagram shows how various sections of the organization work together to define what they want from the feedback

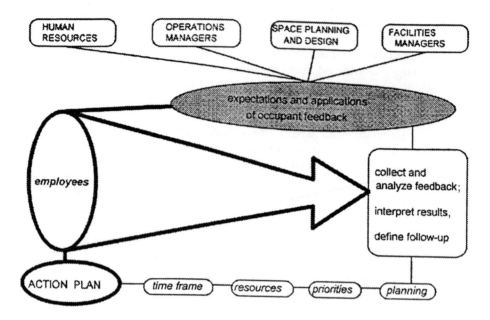

Figure 11.1. Diagram showing how feedback from building occupants is incorporated into a decision-making system.

exercise. Similarly, employees are informed so that their expectations of the results are realistic. Subsequent to these preparatory stages, the feedback is systematically collected from occupants and analyzed. Once the results have been interpreted, various sections of the organization reconvene to decide how they want to use the information: to whom it should be disseminated, what form it should take, and how it should be followed-up. Decisions regarding follow-up require priority-setting on resources, actions by more than one department, and often a wait before funds become available. These are the elements of the action plan, and it is from these elements that occupants are in turn informed of the planning that is going on and the actions that are likely to result. Communication with occupants occurs both at the beginning and the end of each cycle as well as being built in to the cycle itself.

In organizations that already carry out rational facilities planning—proactive planning that is related to organizational goals and predicated on the long-term consumption of resources—systematic feedback from occupants informs and improves decisions. In companies where facilities planning is more of an exercise in avoiding disasters and reducing costs, feedback from occupants functions as an impetus for a more rational planning process.

IMPLEMENTATION OF AN "INTEGRATED WORKPLACE STRATEGY"[15]

The strategic advantage of a planning process designed around feedback from building occupants is that it provides a ready-made steppingstone to the design of an integrated workplace strategy. If a company wants to select the right combination of innovative workspace alternatives to best meet its corporate goals—such as shared offices, telework, and hotelling—and determine how to implement such a plan, a sophisticated accommodation strategy is necessary. This strategy, in combination with human resources planning, technology development, training and education, and business planning, is known as an "Integrated Workplace Strategy." Such a dramatic change to an organization, by definition, cannot succeed if imposed on employees by facilities staff and real estate planners. The integrated planning of space alternatives and business needs, such as expansion or downsizing, is one distinct way in which facilities planning is destined to link into the business units of the organization.

An integrated workplace strategy contains the following elements:

1. careful *fiscal analysis* of costs and benefits to be realized through redefinition of space requirements and implementation of alternatives;

2. *feedback from occupants* on their functional comfort requirements;

3. analysis of *information and communications technology* required to decentralize certain workgroup operations;

4. training for supervisors in strategies and techniques of *long-distance supervision;*

5. preparation and training in *time management*, in company standards, and in the *values and attitude changes* needed at all levels to conform to a new way of working;

6. redefinition of *output and productivity* indicators to conform to geographically dispersed workgroup configurations;

7. redefinition, ultimately, of the *meaning of work* and the meaning of having "a job."

These steps are part of the process of moving corporate environments into the twenty-first century. If planners want alternative work environments to have hoped-for results in terms of increased productivity and lower accommodation costs, feedback from users applied to the design of alternatives is essential. Where workspaces are used interchangeably by individuals, where people can do work from wherever they happen to be, where layouts are designed to conform to work-flow processes, and where space is streamlined to contribute optimally to efficient task performance, feedback from employees on environmental conditions conducive to the optimal and effective performance of work is fundamentally necessary.

The risk of failing to employ a rational planning process for alternative workspaces engenders the same results as the rapid and unplanned introduction of office automation in the 1980s. Companies rushed to purchase computers, especially PC's, with little regard to job definition and the effects on their employees. As Zuboff points out, the result was more often "to automate" or to use the new technology to confirm and enhance the *status quo*, than " to informate," or use the new technology to bring about effective change.[16]

> The absence of a self-conscious strategy to exploit the informating capacity of the new technology has tended to mean that managerial action flows along the path of least resistance—a path that, at least superficially, appears only to serve the interests of managerial hegemony.[17]

To impose innovative work-styles and places on employees without using occupants' feedback risks expending large amounts on confirming the *status quo*. Employees are likely to experience increased anxiety and fears about possible layoffs and lost job status. They fear the increase in power and status of those who remain in a headquarters office building while they move out. Such innovation may even reduce the effectiveness of existing technology and contribute in other ways to lost productivity. At the very least, innovative work environments without organizational change is emphatically a lost opportunity, to the eventual detriment of corporate performance and even survival in the economy of the future. Using the impetus to establish alternative workplaces as an opportunity to plan an integrated workplace strategy, to integrate business and real estate planning, and to empower employees, will have long-term beneficial effects on the organization and will ensure its survival well into the next century.

To be effective in designing alternative work environments for their employees, managers must address and coordinate human resources issues, business strategy, the impact of electronic technology and new equipment, as well as real estate issues and the management of real estate assets. The diagram of the occupants' feedback process (Fig. 11.1) shows how representatives of these same areas of corporate interest and activity convene to plan applications of occupants' feedback, and incorporate communication with the occupants themselves. A similar kind of procedure is necessary to plan alternative work arrangements and environments, but cannot be implemented effectively without the cooperation of the various parts of the organization, nor without the involvement of employees. With reasoned analysis and careful planning, it is possible to reduce the amount of space needed and to ensure at the same time that it is optimal quality space for employees.

FINAL WORDS

In the first chapter, it was pointed out that the O–A relationship for many companies and public agencies often deteriorates over time and ends with an organization deciding to move out of its space. This forecast is as depressing as concluding that most marriages eventually end in divorce. As people grow and change and are altered by their experiences and by their relationships, so the marriage relationship needs to adapt and grow to continue to meet their needs. The office environment is no more fixed and immutable than the institution of marriage: like marriage, there are as many ways of defining it as there are people doing it. Large organizations in downtown high-rise office buildings are learning to shrink space, attract tenants, offer work-at-home and satellite office alternatives, and invest in the best electronic communications equipment available. Small entrepreneurial companies, who are often priced out of the market for downtown office space, are investing in a variety of innovative practices, such as working from home, sharing facilities, and doubling up with other small companies, decentralizing their parts and products inventory, and hiring employees who work at home.[18] In all these cases, employees retain control over the ambient environmental conditions they need for work.

To prepare for the future, companies most likely to be successful are sharing profits with employees, out-sourcing a large number of their support functions, using electronic technology to liberate their employees' workspace from space and time constraints, and taking other actions that are predicated upon employee input and ideas. As Handy points out, "A learning organization needs to have a formal way of asking questions, seeking out theories, testing them, and reflecting on them."[19] And where better to start than the occupants, the employees, the users themselves, whose numbers are growing, who will work harder to keep their jobs, whose performance is ever more important as a result of the amount of training and education invested in them, and whose standards are rising as they become more sophisticated and experienced at office work. On further consideration, it becomes clear that there is no other rationale for office space planning. Simple cost cutting is not a long-term guideline for valid space decisions. Even increasing services to occupants and asking them if they are satisfied is not a sufficient basis for long-term resource allocation and space planning.

What are needed are increased spatial responsibility and awareness at all corporate levels, and these can best be achieved through a planning rationale that systematically takes occupant requirements as the basis for decision-making. Among the influential ideas of Le Corbusier, the great architect, is the notion of the house as a machine for living: a streamlined and efficient environment for completing domestic tasks and carrying out the activities of family life. This notion provided an impetus to the modernist movement in architecture and, eventually, contributed to changes in people's life-styles and family structure. To recall the power of this aphorism now, in the midst of our office space crisis, is to turn it into a useful post-modern dictum for work environments: the office is a machine for work. In this light, its relative value and performance can only be assessed in terms of how well it performs for the people who are doing the work. Taking steps to turn the office into a machine for work that people can use to improve their performance at work is to bring the office of yesterday and today into the twenty-first century.

NOTES AND REFERENCES

1. R.H. Waterman, T.J. Peters, and J.R. Phillips, "The 7-S framework," in J.B. Quinn, H. Mintzberg, and R.M.James, eds. *The Strategy Process: Concepts, Contexts and Cases*, (Englewood Cliffs, N. J.: Prentice-Hall, 1988); pp. 271–276
2. Waterman, Peters, and Phillips, "The 7-S Framework", p. 275.
3. A.C. Hax, and N.S. Majluf, *The Strategy Concept and Process: A Pragmatic Approach*, (Englewood Cliffs, N. J.: Prentice-Hall, 1991) chap. 18.
4. Hax and Majluf, *The Strategy Concept and Process*, p. 27.
5. Hax and Majluf, *The Strategy Concept and Process*, p. 287.
6. Hax and Majluf, *The Strategy Concept and Process*, p. 296
7. Hax and Majluf, *The Strategy Concept and Process*, pp. 302-303.
8. Hax and Majluf, *The Strategy Concept and Process*, chap. 19.
9. Michael E. Porter, *Competitive Advantage: Creating and Sustaining Superior Performance*, (New York: The Free Press, 1985) chap 2.
10. Porter, *Competitive Advantage*, p. 36.
11. Porter, *Competitive Advantage*, p. 38
12. Porter, *Competitive Advantage*, p. 39.
13. Another dramatic example is offered by a music school interested in ethno-musicology, which purchased a gamelan, a very large musical instrument designed for group use and needed a small room to accommodate it. Lack of space on campus meant the instrument was inappropriately accommodated in a library where other students were studying and, therefore, could only be used two hours a day until ap-

propriate space was found. The "technology" was rendered ineffectual by inappropriate accommodation, and the accommodation was inappropriate because of the lack of planning.

14. Senge, *The Fifth Discipline*, p. 78
15. Joroff, Louargand, Lamber, and Becker, *Strategic Management of the Fifth Resource*, p. 54
16. Shoshanna Zuboff, *In the Age of the Smart Machine: The Future of Work and Power*, (New York: Basic Books, 1988). pp. 10–11
17. Zuboff, *In the Age of the Smart Machine*, p. 391.
18. John Case,"A Company of Businesspeople" *INC Magazine*, April 1993. p.79.
19. Handy, *The Age of Unreason* p. 225.

THE BUILDING-IN-USE DATABASES AND HOW THEY ARE USED

There are four BIU databases. They are listed in Table 1.

The proprietary databases are used by companies who want their own norms, against which they can compare building scores. The two BIU databases are used for baseline scores against which to compare other building scores. The norms calculated for the second and more recent BIU databases are shown in Figure 1. Statistical analysis of the original Public Works Canada BIU database datermined that the seven BIU dimensions are stable enough to predict their existence in all North American office buildings. Scores generated on the BIU dimensions in European buildings have also been usefully compared to the database norms. Subsequent analysis of the more recent databases has confirmed that the norms are consistent enough to serve as baseline scores for each of the seven dimensions.

The BIU dimensions were the result of a detailed data analysis using factor analysis techniques and other multivariate statistical analyses. This process is described in the technical report: *Derivation of the Tenant Questionnaire Survey Assessment Method: Office Building Occupant Survey Data Analysis*, published in 1987 by Public Works Canada. The scales used in the questionnaire are reliable and valid predictors of each dimension, and constitute the smallest number of questions that can be

Table 1 BIU Databases

	Content	Size	Building types	Age
1. Original Public Works Canada database	Five Government of Canada buildings, located throughout the country.	2900 cases	Large, modern, multi-story office buildings. (100,000 200,000 square feet)	Built in 1970s and early 1980s assessed mid 1980s
2. Newer, BIU-acquired database	Eight commercial office buildings, located in the North-eastern US and Canada.	3,900 cases	Large and medium-sized, owner-occupied or leased office buildings in city centers and suburbs. (30,000 to 100,000 sq. ft.)	Built or renovated in 1980s; assessed late 1980's and early 1990s.
3. World Bank database (Information Technology and Facilities)	Eight owner-occupied buildings and some leased office space located in Washington, D.C.	2600 cases	Large and medium-sized downtown office buildings built and/or fitted out to World Bank standards.	Buildings of varying ages, but all renovated in early 1990's, assessed in early 1990s.
4. Bell Canada database	Six owned and leased office buildings located in eastern Canada.	2200 cases	Mostly large office buildings in suburbs and downtown.	Owned buildings dating from 70s and 80s; leased space in newer buildings.

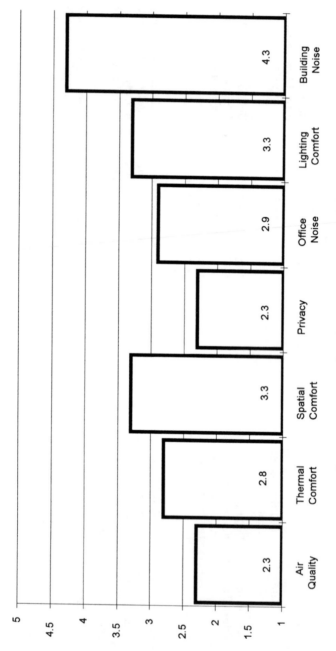

Figure 1. BIU norms based on database calculations.

asked in order to reliably predict the score on each dimension. The scales on which ratings are required for a BIU Assessment are listed in Table 2.

The norms for each database are generated through careful selection of data to be included in each database (atypical buildings or problem buildings are not included in the database), and by examining mean scores for the selected groups of buildings. Analysis of variance indicates those scores or buildings which deviate significantly from the mean score that has been computed, and if these differences cannot be explained, the building is dropped from the database. Figure 2 shows the distribution of all BIU scores in the eight buildings, the second BIU database. The figure demonstrates the consistency of BIU scores across buildings.

The Building-In-Use dimensions are defined as follows, with the normative or baseline score on the 5-point scale in parentheses (first BIU database).

Air Quality (2.3) is a measure of occupants' experience of ventilation conditions, air freshness, and whether the air is circulating or stale. It also indicates whether occupants are aware of odors in the air, whether they feel too warm, and whether the air is dry. *Thermal Comfort* (2.8) is a measure of occupants' experience of cold temperatures, drafts and fluctuating temperatures, as well as general temperature comfort.

Spatial Comfort (3.3) expresses occupants' experience of the spatial layout of the workspace, furniture arrangements, adequacy of personal and work storage, and amount of space people have to do their work. It is closely related to *Privacy* (2.3), which is a separate dimension indicating occupants' experience of conversational privacy, including telephone conversations, and their visual privacy. *Office Noise Control* (2.9) is computed from occupants' ratings of background office noise levels, noise generated by people and nearby equipment, and specific intrusive noises from voices or machines. Good control means a comfortable noise level, whereas too much noise is expressed as a low score indicating discomfort.

Lighting Comfort (3.3) is a measure of occupant's experience of visual comfort, including lack of glare, sufficient light, colors and access to windows and daylight. The seventh Building-In-Use dimension, Building *Noise Control* (4.3), is computed from occupants' ratings of noise emitted from remote or building-related sources, such as mechanical systems, building lights, and traffic or other noise from outside the building. Occupants assess noise from building-related sources differently from noise generated by their coworkers' activities: tolerance for building noise is higher than tolerance for office noise.

Table 2 BIU Assessment Questionnaire

Temperature Comfort	1 GENERALLY BAD	2	3	4	5 GENERALLY GOOD
How Cold It Gets	1 TOO COLD	2	3	4	5 COMFORTABLE
How Warm It Gets	1 TOO WARM	2	3	4	5 COMFORTABLE
Temperature Shifts	1 TOO FREQUENT	2	3	4	5 GENERALLY CONSTANT
Ventilation Comfort	1 GENERALLY BAD	2	3	4	5 GENERALLY GOOD
Air Freshness	1 STALE AIR	2	3	4	5 FRESH AIR
Air Movement	1 STUFFY	2	3	4	5 CIRCULATING
Noise Distractions	1 GENERALLY BAD	2	3	4	5 GENERALLY GOOD
General Office Noise Level (Background Noise from Conversation and Equipment)	1 TOO NOISY	2	3	4	5 COMFORTABLE
Specific Office Noises (Voices and Equipment)	1 DISTURBING	2	3	4	5 NOT A PROBLEM
Noise from the Air Systems	1 DISTURBING	2	3	4	5 NOT A PROBLEM
Noise from the Office Lighting	1 BUZZ/NOISY	2	3	4	5 NOT A PROBLEM

Noise from Outside the Building	1 DISTURBING	2	3	4	5 NOT A PROBLEM
Furniture Arrangement in Your Workspace	1 GENERALLY BAD	2	3	4	5 GENERALLY GOOD
Amount of Space in Your Workspace	1 INSUFFICIENT	2	3	4	5 ADEQUATE
Work Storage	1 INSUFFICIENT	2	3	4	5 ADEQUATE
Personal Storage	1 INSUFFICIENT	2	3	4	5 ADEQUATE
Visual Privacy at Your Desk	1 BAD	2	3	4	5 GOOD
Voice Privacy at Your Desk	1 BAD	2	3	4	5 GOOD
Telephone Privacy at Your Desk	1 BAD	2	3	4	5 GOOD
Electrical Lighting	1 BAD	2	3	4	5 GOOD
How Bright Lights Are	1 TOO MUCH LIGHT	2	3	4	5 DOES NOT GET TOO BRIGHT
Glare from Lights	1 HIGHT GLARE	2	3	4	5 NO GLARE

Please rate whether or not this space helps you do your work?

	1 MAKES IT DIFFICULT	2	3	4	5 MAKES IT EASY

How would you rate your overall satisfaction with this building?

	1 DISSATISFIED	2	3	4	5 VERY SATISFIED

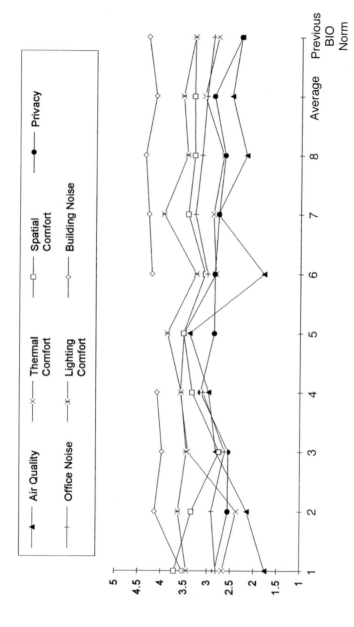

Figure 2. Range of BIU scores received for all buildings in the current BIU database.

In order to determine whether or not the deviation of a score from the norm is significant, a T-test of the significance of the difference between two means is applied to the norm and the score for that sample size (or for the whole building). As a result of the test, the score for each dimension relative to the normative score can be labeled: "Very good, good, fair, average, below average, poor, very poor." Users are then encouraged to relate the results to conditions in their building in order to interpret them meaningfully enough to allow effective follow-up action.

Building-In-Use Profile results can be further analyzed according to location differences, depending on where respondents sit in the building. The primary location category is floor: responses for each floor are analyzed separately and are compared to the scores received from the whole building in order to determine the good and bad comfort conditions on each floor. In addition, differences between open and enclosed office, and between window and interior locations can be analyzed. However, it is important to understand what is sought from this degree of detail in order to make the analysis worthwhile. BIU differences among workgroups—if group members are locationally contiguous and if the type of work being performed by different groups is markedly different—can also be examined. It is possible to examine separate air handling zones for variation in air quality and thermal comfort scores, perimeter and interior zones for variation in lighting comfort and thermal comfort scores, and enclosed and open office for variation in privacy and noise control scores.

It is important to emphasize that this stage of data analysis is not seeking to prove or disprove hypothetical relationships between location attributes and BIU scores. The purpose of the detailed stage of data analysis is to link BIU scores more closely to actual building conditions. A research approach would seek consistency across buildings, so that, for example, an index score of -0.2 on office noise control can reliably be predicted from office workers in open plan workstations in all buildings. The BIU system, being an action approach, looks at each building independently, and, finding that in an open plan building, floor 4 has an office noise control index of -0.2 and floor 5 has an index of 0.1, infers from conditions observed in the building that a well-used employee lounge on floor 4 is creating additional noise problems for occupants that cause more discomfort than the open plan layout per se. This process of relating BIU scores to real building conditions is called *Interpretation of Results*. The process allows a problem-solving action to be inferred: namely, lounge access controlled, a higher standard of party wall construction applied, or a new office floorplan designed that separates the lounge from areas where people are working.

It is more useful to compare individual floor scores to its building's profile than to the BIU norms, as each building's profile is unique and the differences between floor scores and profile scores will be relatively small. Interpreting results requires being sensitive to small differences in building conditions because the action recommendations that are the product of the exercise are designed to address the process of overall improvement to an existing space, even one that is already of a high standard. They are designed to become part of an ongoing facilities workplan of maintenance, renovation, and repair rather than to add expensive additional items to the facilities budget. Although some element of problem correction is involved, major and glaring problems, such as threats to worker health, manifest themselves elsewhere than in occupant surveys, and are better addressed through larger and more expensive interventions than are generally the product of Building-In-Use Assessment.

INDEX